KSA
Grade 6 Math Practice

GET DIGITAL ACCESS TO

2 KSA Practice Tests

Personalized Study Plans

REGISTER NOW

Link

QR Code

Visit the link below for online registration

lumoslearning.com/a/tedbooks

Access Code: KSAG6M-40619-P

Kentucky Summative Assessment Test Prep: 6th Grade Math Practice Workbook and Full-length Online Assessments: KSA Study Guide

Contributing Editor - Renee Bade
Contributing Editor - Kimberly G.
Executive Producer - Mukunda Krishnaswamy
Program Director - Anirudh Agarwal
Designer and Illustrator - Sowmya R.

ISBN 13: 978-1959697671

Printed in the United States of America

FOR SCHOOL EDITION AND PERMISSIONS, CONTACT US

LUMOS INFORMATION SERVICES, LLC

PO Box 1575, Piscataway, NJ 08855-1575
www.LumosLearning.com

Email: support@lumoslearning.com
Tel: (732) 384-0146
Fax: (866) 283-6471

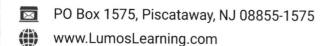

Lumos Learning
Step Up Your Skills

INTRODUCTION

This book is specifically designed to improve student achievement on the Kentucky Summative Assessment (KSA). Students perform at their best on standardized tests when they feel comfortable with the test cotent as well as the test format. Lumos online practice tests are meticulously designed to mirror the state assessment. They adhere to the guidelines provided by the state for the number of sessions and questions, standards, difficulty level, question types, test duration and more.

Based on our decade of experience developing practice resources for standardized tests, we've created a dynamic system, the Lumos Smart Test Prep Methodology. It provides students with realistic assessment rehearsal and an efficient pathway to overcoming each proficiency gap.

Use the Lumos Smart Test Prep Methodology to achieve a high score on the KSA.

Lumos Smart Test Prep Methodology

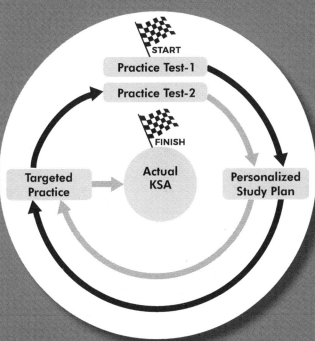

1 The student takes the first online diagnostic test, which assesses proficiency levels in various standards.

2 StepUp generates a personalized online study plan based on the student's performance.

3 The student completes targeted practice in the printed workbook and marks it as complete in the online study plan.

4 The student then attempts the second online practice test.

5 StepUp generates a second individualized online study plan.

6 The student completes the targeted practice and is ready for the actual KSA.

Table of Contents

Chapter 1
Ratios & Proportional Relationships

Lesson 1: Expressing Ratios

1. **A school has an enrollment of 600 students. 330 of the students are girls. Express the fraction of students who are boys in simplest terms.**

   Ⓐ $\dfrac{12}{20}$

   Ⓑ $\dfrac{11}{20}$

   Ⓒ $\dfrac{9}{20}$

   Ⓓ $\dfrac{13}{20}$

2. **In the 14th century, the Sultan of Brunei noticed that his ratio of emeralds to rubies was the same as the ratio of diamonds to pearls. If he had 85 emeralds, 119 rubies, and 45 diamonds, how many pearls did he have?**

   Ⓐ 17
   Ⓑ 22
   Ⓒ 58
   Ⓓ 63

3. **Mr. Fullingham has 75 geese and 125 turkeys. What is the ratio of the number of geese to the total number of birds in simplest terms?**

   Ⓐ 75:200
   Ⓑ 3:8
   Ⓒ 125:200
   Ⓓ 5:8

4. **The little league team called the Hawks has 7 brunettes, 5 blonds, and 2 redheads. What is the ratio of redheads to the entire team in simplest terms?**

   Ⓐ 2:7
   Ⓑ 2:5
   Ⓒ 2:12
   Ⓓ 1:7

5. The little league team called the Hawks has 7 brunettes, 5 blonds, and 2 redheads. The entire little league division that the Hawks belong to has the same ratio of redheads to everyone else. What is the total number of redheads in that division if the total number of players is 126?

   Ⓐ 9
   Ⓑ 14
   Ⓒ 18
   Ⓓ 24

6. Barnaby decided to count the number of ducks and geese flying south for the winter. The first day he counted 175 ducks and 63 geese. What is the ratio of ducks to the total number of birds flying overhead in simplest terms?

   Ⓐ 175:63
   Ⓑ 175:238
   Ⓒ 25:9
   Ⓓ 25:34

7. Barnaby decided to count the number of ducks and geese flying south for the winter. The first day he counted 175 ducks and 63 geese. By the end of migration, Barnaby had counted 4,725 geese. If the ratio of ducks to geese remained the same (175 to 63), how many ducks did he count?

   Ⓐ 13,125
   Ⓑ 17,850
   Ⓒ 10,695
   Ⓓ 14,750

8. Barbara was baking a cake and could not find her tablespoon measure. The recipe calls for $3\frac{1}{3}$ tablespoons. Each table spoon measure 3 teaspoon. How many teaspoons must Barbara use in order to have the recipe turn out all right?

   Ⓐ 3
   Ⓑ 6
   Ⓒ 9
   Ⓓ 10

9. The ratio of girls to boys in a grade is 6 to 5. If there are 24 girls in the grade then how many students are there altogether?

   Ⓐ 14
   Ⓑ 24
   Ⓒ 34
   Ⓓ 44

10. The ratio of pencils to pens in a box is 3 to 2. If there are 30 pencils and pens altogether, how many pencils are there?

    Ⓐ 16
    Ⓑ 17
    Ⓒ 18
    Ⓓ 19

11. Which of the following correctly expresses the ratio of shaded bows to the number of total bows? Select all answers that apply.

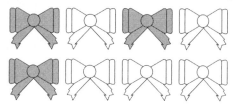

    Ⓐ 3:8
    Ⓑ 5:8
    Ⓒ $\dfrac{3}{5}$
    Ⓓ $\dfrac{3}{8}$
    Ⓔ $\dfrac{5}{8}$

12. Write the ratio that correctly describes the number of white stars compared to the number of gray stars. Write your answer in the box below.

13. Complete the following table by filling in the blanks with a number that shows the correct ratio that is equivalent to the one shown in the first row.

1	2
2	4
	6
4	8
5	
	12

CHAPTER 1 → Lesson 2: Unit Rates

1. **Which is a better price: 5 for $1.00, 4 for 85¢, 2 for 25¢, or 6 for $1.10?**

   Ⓐ 5 for $1.00
   Ⓑ 4 for 85¢
   Ⓒ 2 for 25¢
   Ⓓ 6 for $1.10

2. **At grocery Store A, 5 cans of baked beans cost $3.45. At grocery Store B, 7 cans of baked beans cost $5.15. At grocery Store C, 4 cans of baked beans cost $2.46. At grocery Store D, 6 cans of baked beans cost $4.00. How much money would you save if you bought 20 cans of baked beans from grocery store C than if you bought 20 cans of baked beans from grocery store A?**

   Ⓐ $1.75
   Ⓑ $1.25
   Ⓒ $1.50
   Ⓓ 95¢

3. **Beverly drove from Atlantic City to Newark. She drove for 284 miles at a constant speed of 58 mph. How long did it take Beverly to complete the trip?**

   Ⓐ 4 hours and 45 minutes
   Ⓑ 4 hours and 54 minutes
   Ⓒ 4 hours and 8 minutes
   Ⓓ 4 hours and 89 minutes

4. **Don has two jobs. For Job 1, he earns $7.55 an hour. For Job 2, he earns $8.45 an hour. Last week he worked at the first job for 10 hours and at the second job for 15 hours. What were his average earnings per hour?**

   Ⓐ $8.00
   Ⓑ $8.09
   Ⓒ $8.15
   Ⓓ $8.13

5. **It took Marjorie 15 minutes to drive from her house to her daughter's school. If the school was 4 miles away from her house, what was her unit rate of speed?**

   Ⓐ 16 mph
   Ⓑ 8 mph
   Ⓒ 4 mph
   Ⓓ 30 mph

6. The Belmont race track known as "Big Sandy" is 1½ miles long. In 1973, Secretariat won the Belmont Stakes race in 2 minutes and 30 seconds. Assuming he ran on "Big Sandy", what was his unit speed?

    Ⓐ 30 mph
    Ⓑ 40 mph
    Ⓒ 36 mph
    Ⓓ 38 mph

7. If 1 pound of chocolate creams at Philadelphia Candies costs $7.52. How much does that candy cost by the ounce?

    Ⓐ 48¢ per oz.
    Ⓑ 47¢ per oz.
    Ⓒ 75.2¢ per oz.
    Ⓓ 66¢ per oz.

8. If Carol pays $62.90 to fill the 17-gallon gas tank in her vehicle and she can drive 330 miles on one tank of gas, about how much does she pay per mile to drive her vehicle?

    Ⓐ $0.37
    Ⓑ $3.70
    Ⓒ $0.19
    Ⓓ $0.01

9. A 13 ounce box of cereal costs $3.99. What is the unit price per pound?

    Ⓐ about $1.23
    Ⓑ about $2.66
    Ⓒ about $4.30
    Ⓓ about $4.91

10. A bottle of perfume costs $26.00 for a $\frac{1}{2}$ ounce bottle. What is the price per ounce?

    Ⓐ $25.50
    Ⓑ $26.50
    Ⓒ $52.00
    Ⓓ $13.00

11. Check the box in each row that represents the correct unit rate for each situation.

	1:10	1:5	1:50	1:20
$1.00 per 5 pounds	◯	◯	◯	◯
50 pounds per box	◯	◯	◯	◯
10 miles per gallon	◯	◯	◯	◯
1 lap in 20 minutes	◯	◯	◯	◯

12. Below is a recipe for Grandma Grittle's favorite cupcakes. Which of the following correctly expresses a ratio found in the recipe? Check all answers that apply.

Ingredients for 12 cupcakes:
White flour - 2 cups
Sugar - 1 cup
Baking powder - 2 tsp.
Salt - 1 tsp.
Butter or margarine - 1/3 cup
Milk - 2/3 cup
Vanilla - 1 tsp.
Semi-sweet chocolate - 1 bar

Ⓐ sugar and flour, 1:2
Ⓑ flour and total cupcakes, 1:12
Ⓒ vanilla and salt, 1:1
Ⓓ salt and baking powder, 1:2
Ⓔ chocolate to total cupcakes, 1:24

13. The bag of apples shown in the picture, costs $3.20. The cost of one apple is _____ .

CHAPTER 1 → Lesson 3: Solving Real World Ratio Problems

1. **How many kilograms are there in 375 grams?**

   Ⓐ 3,750 kg
   Ⓑ 37.5 kg
   Ⓒ 3.75 kg
   Ⓓ 0.375 kg

2. **How many inches are there in 2 yards?**

   Ⓐ 24 in
   Ⓑ 36 in
   Ⓒ 48 in
   Ⓓ 72 in

3. **What is 50% of 120?**

   Ⓐ 50
   Ⓑ 60
   Ⓒ 70
   Ⓓ 55

4. **Michael Jordan is six feet 6 inches tall. How much is that in inches?**

   Ⓐ 66 inches
   Ⓑ 76 inches
   Ⓒ 86 inches
   Ⓓ 78 inches

5. **What is 7.5% in decimal notation?**

   Ⓐ 0.75
   Ⓑ 0.075
   Ⓒ 0.0075
   Ⓓ 7.5

6. **A $60 shirt is on sale for 30% off. How much is the shirt's sale price?**

   Ⓐ $30
   Ⓑ $40
   Ⓒ $18
   Ⓓ $42

7. On Monday, 6 out of every 10 people who entered a store purchased something. If 1,000 people entered the store on Monday, how many people purchased something?

   Ⓐ 6 people
   Ⓑ 60 people
   Ⓒ 600 people
   Ⓓ 610 people

8. If a pair of pants that normally sells for $51.00 is now on sale for $34.00, by what percentage was the price reduced?

   Ⓐ 30%
   Ⓑ 60%
   Ⓒ 33.33%
   Ⓓ 66.67%

9. If Comic Book World is taking 28% off the comic books that normally sell for $4.00, how much money is Kevin saving if he buys 12 comic books during the sale?

   Ⓐ $28
   Ⓑ $12
   Ⓒ $13.44
   Ⓓ $14.58

10. Eric spends 45 minutes getting to work and 45 minutes returning home. What percent of the day does Eric spend commuting?

    Ⓐ 6.25%
    Ⓑ 7.8%
    Ⓒ 5.95%
    Ⓓ 15%

11. 75% of the crowd at a sports rally were wearing the team colors. If 222 people were wearing the team colors, how many people were in the crowd? Circle the correct answer choice.

    Ⓐ 74
    Ⓑ 300
    Ⓒ 296
    Ⓓ 314

12. To make yummy fruit punch, use 2 cups of grape juice for every 3 cups of apple juice. Select all of the juice combinations below that correctly follow this recipe ratio.

Ⓐ 4 cups grape juice: 6 cups apple juice
Ⓑ 5 cups grape juice: 10 cups apple juice
Ⓒ 6 cups grape juice: 9 cups apple juice
Ⓓ 6 cups grape juice: 12 cups apple juice
Ⓔ All of the above

13. Look at the ratio information found in the table below. Complete the table by correctly filling in the missing information.

Feet	Yards
3	1
6	
	3
15	5
24	

CHAPTER 1 → Lesson 4: Solving Unit Rate Problems

1. **A 12 pack of juice pouches costs $6.00. How much does one juice pouch cost?**

   Ⓐ $0.02
   Ⓑ $0.20
   Ⓒ $0.50
   Ⓓ $0.72

2. **Eli can ride his scooter 128 miles on one tank of gas. If the scooter has a 4-gallon gas tank, how far can Eli ride on one gallon of gas?**

   Ⓐ 64 miles
   Ⓑ 32 miles
   Ⓒ 512 miles
   Ⓓ 20 miles

3. **Clifton ran 6 miles in 39 minutes. At this rate, how much time Clifton takes to run one mile?**

   Ⓐ 13 minutes
   Ⓑ 12 minutes
   Ⓒ 7.2 minutes
   Ⓓ 6 minutes and 30 seconds

4. **Brad has swimming practice 3 days a week. This week Brad swam a total of 114 laps. At this rate how many laps did Brad swim each day?**

   Ⓐ 38 laps
   Ⓑ 42 laps
   Ⓒ 57 laps
   Ⓓ 61 laps

5. **Karen bought a total of seven items at five different stores. She began with $65.00 and had $15.00 remaining. Which of the following equation can be used to determine the average cost per item?**

   Ⓐ $7x \times 5 = \$50.00$
   Ⓑ $7x = \$75.00$
   Ⓒ $7x + \$15.00 = \$65.00$
   Ⓓ $5x = \$65.00 - \$15.00$

6. Geoff goes to the archery range five days a week. He must pay $1.00 for every ten arrows that he shoots. If he spent $15.00 this week on arrows what is the average number of arrows Geoff shot per day?

   (A) 3 arrows
   (B) 30 arrows
   (C) 45 arrows
   (D) 75 arrows

7. Julia made 7 batches of cookies and ate 3 cookies. There were 74 cookies left. Which expression can be used to determine the average number of cookies per batch?

   (A) $74 \div 7$
   (B) $(74+7) \div 3$
   (C) $\dfrac{74 + 3}{7}$
   (D) $\dfrac{74}{3} \times 7$

8. Lars delivered 124 papers in 3 hours. How long did it take Lars to deliver one paper?

   (A) 1 minute
   (B) 1 minute and 27 seconds
   (C) 1 minute and 45 seconds
   (D) 2 minutes and 3 seconds

9. Mr. and Mrs. Fink met their son Conrad at the beach. Mr. and Mrs. Fink drove 462 miles on 21 gallons of fuel. Conrad drove 456 miles on 12 gallons of fuel. How many more miles per gallon does Conrad's car get than Mr. and Mrs. Fink's car?

   (A) 6 mpg
   (B) 22 mpg
   (C) 16 mpg
   (D) 38 mpg

10. Myka bought a box of 30 greeting cards for $4.00. Chuck bought a box of 100 greeting cards for $12.00. Who got the better deal?

    (A) Myka got the better deal at about 13 cents per card.
    (B) Myka got the better deal at about 7.5 cents per card.
    (C) Chuck got the better deal at about 8 cents per card.
    (D) Chuck got the better deal at 12 cents per card.

11. **John paid $15 for 3 cheeseburgers. What is the rate of one cheeseburger? Enter your answer in the box below.**

$

12. **Tommy charges the same rate for each yard he mows. Calculate the rate he charges, then complete the missing information in the table.**

Day	Total Money Earned $	Yards Mowed
Monday	50	2
Wednesday		3
Friday	25	
Saturday		5

13. **A movie theatre charges $10 for a ticket. Check the box in each row that represents how much money the theatre would make from ticket sales.**

	$150	$10	$100	$200
10 tickets	☐	☐	☐	☐
15 tickets	☐	☐	☐	☐
20 tickets	☐	☐	☐	☐
1 ticket	☐	☐	☐	☐

CHAPTER 1 → Lesson 5: Finding Percent

1. **What is 25% of 24?**

   Ⓐ 5
   Ⓑ 6
   Ⓒ 11
   Ⓓ 17

2. **What is 15% of 60?**

   Ⓐ 9
   Ⓑ 12
   Ⓒ 15
   Ⓓ 25

3. **9 is what percent of 72?**

   Ⓐ 7.2%
   Ⓑ 8%
   Ⓒ 12.5%
   Ⓓ 14%

4. **How much is 30% of 190?**

   Ⓐ 45
   Ⓑ 57
   Ⓒ 60
   Ⓓ 63

5. **Daniel has 280 baseball cards. 15% of these are highly collectable. How many baseball cards does Daniel possess that are highly collectable?**

   Ⓐ 15 cards
   Ⓑ 19 cards
   Ⓒ 42 cards
   Ⓓ 47 cards

6. **The football team consumed 80% of the water provided at the game. If the team consumed 8-gallons of water, how much water was provided?**

   Ⓐ 10 gallons
   Ⓑ 12 gallons
   Ⓒ 15 gallons
   Ⓓ 18.75 gallons

7. Joshua brought 156 of his 678 Legos to Emily's house. What percentage of his Legos did Joshua bring?

    Ⓐ 4%
    Ⓑ 23%
    Ⓒ 30%
    Ⓓ 43%

8. At batting practice Alexis hit 8 balls out of 15 into the outfield. Which equation below can be used to determine the percentage of balls hit into the outfield?

    Ⓐ $\dfrac{15}{8} = \dfrac{x}{100}$

    Ⓑ $\dfrac{15}{100} = \dfrac{x}{8}$

    Ⓒ $8x = (100)(15)$

    Ⓓ $\dfrac{15}{8} = \dfrac{100}{x}$

9. Nikki grows roses, tulips, and carnations. She has 78 flowers of which 32% are roses. Approximately how many roses does Nikki have?

    Ⓐ 18 roses
    Ⓑ 25 roses
    Ⓒ 28 roses
    Ⓓ 41 roses

10. Victor took out 30% of his construction paper. Of this, Paul used 6 sheets, Allison used 8 sheets and Victor and Gayle used the last ten sheets. How many sheets of construction paper did Victor not take out?

    Ⓐ 24 sheets
    Ⓑ 50 sheets
    Ⓒ 56 sheets
    Ⓓ 80 sheets

11. The following items were bought on sale. Complete the missing information.

Item Purchased	Original Price	Amount of Discount	Amount Paid
Video Game	$80	20%	
Movie Ticket	$14		$11.20
Laptop	$1,000		$750
Shoes	$55.00	10%	$49.5

12. Which of these represent 25 percent of the beginning number? Select all that apply.

Ⓐ $90, $22.50
Ⓑ $150, $15.00
Ⓒ $400, $300.00
Ⓓ $560.00, $140.00

CHAPTER 1 → Lesson 6: Measurement Conversion

1. **Owen is 69 inches tall. How tall is Owen in feet?**

    Ⓐ 5.2 feet
    Ⓑ 5.75 feet
    Ⓒ 5.9 feet
    Ⓓ 6 feet

2. **What is 7 gallons 3 quarts expressed as quarts?**

    Ⓐ 4.75 quarts
    Ⓑ 28 quarts
    Ⓒ 29.2 quarts
    Ⓓ 31 quarts

3. **How many centimeters in 3.7 kilometers?**

    Ⓐ 0.000037 cm
    Ⓑ 0.037 cm
    Ⓒ 3700 cm
    Ⓓ 370,000 cm

4. **136 ounces is how many pounds?**

    Ⓐ 6.8 pounds
    Ⓑ 8.5 pounds
    Ⓒ 1088 pounds
    Ⓓ 2,176 pounds

5. **How many ounces in 5 gallons?**

    Ⓐ 128 ounces
    Ⓑ 320 ounces
    Ⓒ 640 ounces
    Ⓓ 1280 ounces

6. **Lisa, Susan, and Chris participated in a three-person relay team. Lisa ran 1284 meters, Susan ran 1635 meters and Chris ran 1473 meters. How long was the race in kilometers? Round your answer to the nearest tenth.**

    Ⓐ 4.0 km
    Ⓑ 4.4 km
    Ⓒ 43.9 km
    Ⓓ 49.0 km

7. Quita recorded the amount of time it took her to complete her chores each week for a month; 1 hour 3 minutes, 1 hour 18 minutes, 55 minutes, and 68 minutes. How many hours did Quita spend doing chores during the month?

   Ⓐ 3.8 hours
   Ⓑ 4.24 hours
   Ⓒ 4.4 hours
   Ⓓ 5.7 hours

8. Lamar can run 3 miles in 18 minutes. At this rate, how much distance he can run in one hour?

   Ⓐ 0.9 miles
   Ⓑ 1.1 miles
   Ⓒ 10 miles
   Ⓓ 21 miles

9. A rectangular garden has a width of 67 inches and a length of 92 inches. What is the perimeter of the garden in feet?

   Ⓐ 13.25 feet
   Ⓑ 26.5 feet
   Ⓒ 31.8 feet
   Ⓓ 42.8 feet

10. Pat has a pen pal in England. When Pat asked how tall his pen pal was he replied, 1.27 meters. If 1 inch is 2.54 cm, how tall is Pat's pen pal in feet and inches?

   Ⓐ 3 feet 11 inches
   Ⓑ 4 feet 2 inches
   Ⓒ 4 feet 6 inches
   Ⓓ 5 feet exactly

11. How many fluid ounces are there in a cup? Circle the correct answer choice.

   Ⓐ 10 fl oz.
   Ⓑ 8 fl oz.
   Ⓒ 4 fl oz.
   Ⓓ 16 fl oz.

12. How many meters are there in 16 kilometers? Circle the correct answer choice.

Ⓐ 1.6 m
Ⓑ 1,600 m
Ⓒ 16,000 m
Ⓓ 160 m

13. Use the chart provided to fill in the missing values in the table below.

1 L	1000 ml
1 g	1000 mg
1 m	1000 mm

3	L		ml
	g	5000	mg
	m	8000	mm
12	L		ml
20	g		mg

End of Ratios & Proportional Relationships

Chapter 1
Ratios & Proportional Relationships
Lesson 1: Expressing Ratios

Question No.	Answer	Detailed Explanations
1	C	First, to find the proper ratio, subtract the number of girls from the total number of students. The difference is the number of boys. $600-330 = 270$. So, the initial ratio is $\frac{270}{600}$. Then, to rewrite a ratio in its simplest terms, divide the numerator and denominator by the Greatest Common Factor (GCF). Here, the GCF is 30. 270 divided by 30 = 9 and 600 divided by 30 = 20, so, the simplest ratio is $\frac{9}{20}$.
2	D	First, find the ratio of emeralds to rubies. That ratio is $\frac{85}{119}$. To find how many pearls the sultan had, set up a proportion with the ratio of diamonds to pearls: $\frac{85}{119} = \frac{45}{x}$ Then, find the cross products of each: $85*x = 119*45$ Simplify: $85x = 5355$ Solve for x by dividing by 85 on both sides: $\frac{85x}{85} = \frac{5355}{85}$ $x = 63$
3	B	$75 + 125 = 200$. Therefore, the total number of birds is 200. The ratio of geese to total birds is 75:200. Simplify the ratio by dividing by the GCF (75,200)= 25, simplified ratio is 3:8.
4	D	There are $(7+5+2) = 14$ players in all. The ratio of redheads to the team is 2:14. Divide by the GCF of 2 to simplify the ratio to 1:7
5	C	Set up the proportion: $\frac{2}{14}=\frac{x}{126}$, $\frac{1}{7}=\frac{x}{126}$, cross multiply to get $7x = 126$, then divide by 7 and $x = 18$.
6	D	The total number of birds is $175+63 = 238$. Thus, the ratio of ducks to total birds is 175:238. To find the ratio in simplest terms, divide by the GCF(175, 238) =7. The ratio in simplest terms is 25:34.

Question No.	Answer	Detailed Explanations

7 — **A**

The ratio of ducks to geese is 175:63. To find how many ducks, set up a proportion of $\frac{175}{63} = \frac{x}{4,725}$.
Find the cross products:
175*4,725 = 63*x
826,875 = 63x
Divide both sides by 63
x = 13,125

8 — **D**

There are 3 teaspoons to each tablespoon. Thus $3 * \frac{10}{3} = 10$ teaspoons.

9 — **D**

To find how many students there are in the grade, set up the proportion $\frac{6}{5} = \frac{24}{x}$.
Notice that you can multiply $\frac{6}{5}$ by $\frac{4}{4}$ to make the numerator of 24. This makes the equivalent denominator 20.
Add 24 + 20 to get the total number of students, or 44.

10 — **C**

If the ratio of pencils to pens is $\frac{3}{2}$ then the ratio of pencils to pencils and pens is $\frac{3}{5}$. To find the number of pencils in a box with 30 pencils and pens, set up the proportion $\frac{3}{5} = \frac{x}{30}$. Then, multiply the first ratio by $\frac{6}{6}$ which will equal $\frac{18}{30}$. There are 18 pencils in the box.

11 — **A & D**

Correct Response: 3:8 and 3/8 There are 3 shaded bows and a total of 8 bows in all. 3:8 is the correct way to write a ratio or it can be written $\frac{3}{8}$, which is read "3 out of 8".

12 — **4:5**

4:5. There are 4 white stars and 5 gray stars.

13

1	2
2	4
3	6
4	8
5	10
6	12

The numbers are 3, 6, 10. The first row shows the ratio pattern, which is 1:2, which means each number in the left column is ½ of the number in the right column.

Lesson 2: Unit Rates

Question No.	Answer	Detailed Explanations
1	C	$\frac{1}{5}$ = a unit price of $0.20 per piece $\frac{.85}{4}$ = a unit price of $0.2125 per piece $\frac{.25}{2}$ = a unit price of $0.125 per piece. This is the best price per unit. $\frac{1.1}{6}$ = a unit price of $0.183 per piece.
2	C	The unit rate at Store A is $\frac{\$3.45}{5}$=$0.69. 20 cans of beans would be $0.69*20= $13.80 The unit rate at Store C is $\frac{\$2.46}{4}$= $0.615. 20 cans of beans would be $0.615*20=$12.30. Subtract $13.80−12.30=$1.50
3	B	284 miles divided by 58 miles per hour are how you will find how long it took Beverly to make the trip. (Distance ÷ rate = time) $\frac{284}{58} \approx 4.9$ hours 0.9 hours = 54 minutes (Multiply 60 by 0.9, because there are 60 minutes in an hour.) 4 hours and 54 minutes is how long it took Beverly to make the trip.
4	B	$7.55 x 10 = $75.50 $8.45 x 15 = $126.75 126.75 + 75.50 = 202.25 $\frac{202.25}{25}$ = $8.09
5	A	$\frac{15}{4}=\frac{60}{x}$, where 60 equals the number of minutes in an hour. 15 x 4 = 60, so multiply the original ratio $\frac{15}{4}$ by $\frac{4}{4}$ to get $\frac{60}{16}$, where 16 represents the miles per hour (mph) that she traveled.
6	C	Set up a ratio of distance/time. Here, the ratio would be $\frac{1.5}{2.5}$ Then, create a proportion $\frac{1.5}{2.5} = \frac{x}{60}$, where 60 represents the number of minutes in an hour. Find the cross products: 1.5*60 = 2.5*x Simplify: 90 = 2.5x, Divide each side by 2.5 we get, x = 36.
7	B	There are 16 ounces in a pound, so $\frac{\$7.52}{16}$ = 47¢
8	C	To find the cost of gas per mile: $\frac{\$62.90}{330}$ equals about $0.19 per mile. (Note: The capacity of the tank is extra information.)

Question No.	Answer	Detailed Explanations
9	D	$3.99/13 equals about $0.306 per ounce. Since there are 16 oz in a pound, multiply 16 by $0.306..., which equals about $4.91.
10	C	$26.00 ÷ (1/2) = $26.00 x 2 = $52.00 per ounce

11

	1:10	1:5	1:50	1:20
$1.00 per 5 pounds		◯		
50 pounds per box			◯	
10 miles per gallon	◯			
1 lap in 20 minutes				◯

Correct Response: A. 1:5, One dollar is spent for every 5 pounds. B. 1:50, Per box refers to a quantity of 1 box, so there are 50 pounds in one box. C. 1:10, Per gallon refers to a quantity of 1 gallon, so every gallon supplies 10 miles. D. 1:20, It takes 20 minutes to run 1 lap.

Question No.	Answer	Detailed Explanations
12	A, C & D	Choices (A), (C) and (D) are all correct. Choice (B) is incorrect because it takes 2 cups of flour to make 12 cupcakes. Choice (E) is incorrect because 1 bar of chocolate is needed for 12 cupcakes, so 2 bars of chocolate would be needed to make 24 cupcakes.
13	$ 0.40	The bag shown in the picture consists of 8 apples. To determine the unit rate for the cost of one apple, divide the cost of all the apples by the total number of apples: $3.20 ÷ 8 = .40 So, one apple has a unit rate cost of $.40.

Lesson 3: Solving Real World Ratio Problems

Question No.	Answer	Detailed Explanations
1	D	1000 grams/1 kilogram = 375 grams/x kilograms 1000x = 375 Divide each side by 1000 x = 0.375 kilograms
2	D	36 inches equal 1 yard, so 72 inches must equal 2 yards.
3	B	is/of = %/100 so: $\frac{x}{120} = \frac{50}{100}$ 100*x = 120*50 100x = 6000 Divide both sides by 100 x = 60
4	D	Since every foot = 12 inches, then 6 feet must equal 72 inches (6*12). Add extra 6 inches to 72 inches which is equal to 78 inches.
5	B	Divide a percentage by 100 to make an equivalent decimal form 7.5/100 = .075
6	D	is/of = %/100 $\frac{x}{60} = \frac{30}{100}$ x*100 = 60*30 100x = 1800 Divide both sides by 100 x = $18 Subtract $18 from $60. $60−$18 = $42
7	C	$\frac{6}{10} = \frac{x}{1000}$ x*10 = 6*1000 10x = 6000 Divide both sides by 10 x = 600
8	C	is/of = %/100 $\frac{34.00}{51.00} = \frac{x}{100}$ 34.00*100 = 51*x 3400 = 51x Divide both sides by 51 x = 66.67% This is the amount left to pay. 100% − 66.67% = 33.33% This is the amount the shirt was reduced by.

Question No.	Answer	Detailed Explanations
9	C	is/of = %/100 $\dfrac{x}{\$4.00} = \dfrac{28}{100}$ 4*28 = 100*x 112 = 100x Divide both sides by 100 x = \$1.12 Then, multiply \$1.12 * 12 = \$13.44
10	A	is/of = x/100 use hours as your proportional rate 45 minutes + 45 minutes = 90 minutes or 1.5 hours $\dfrac{1.5}{24} = \dfrac{x}{100}$ 1.5*100 = 24*x 150 = 24x Divide both sides by 24 x = 6.25%
11	C	$\dfrac{is}{of} = \dfrac{\%}{100}$ $\dfrac{222}{x} = \dfrac{75}{100}$ 222 x 100 = 75 x x 22200 = 75x Divide both sides by 75 **x = 296**
12	A & C	A. $\dfrac{4}{6} = \dfrac{2}{3}$ & C. $\dfrac{6}{9} = \dfrac{2}{3}$ The ratio of grape to apple is 2:3 or $\dfrac{2}{3}$. By fractions equivalent to $\dfrac{2}{3}$, you can determine the correct ratio for the recipe.

13			

Feet	Yards
3	1
6	2
9	3
15	5
24	8

1 yard = 3 feet
Set up the proportion: yard/feet
(1) Let x be the number of yards in 6 feet.
$\dfrac{1}{3} = \dfrac{x}{6}$
3x = 1 x 6 = 6
x = $\dfrac{6}{3}$ = 2 yards
(2) Let y be the number of feet in 3 yards
$\dfrac{1}{3} = \dfrac{3}{y}$
1 x y = 3 x 3 = 9 or y = 9
(3) Let z be the number of yards in 24 feet.
$\dfrac{1}{3} = \dfrac{z}{24}$
3z = 1 x 24 = 24
z = $\dfrac{24}{3}$ = 8

Lesson 4: Solving Unit Rate Problems

Question No.	Answer	Detailed Explanations
1	C	Find the unit rate for one juice pouch. $\dfrac{\$6.00}{12} = \dfrac{x}{1}$ 6*1=12*x 6 = 12x Divide both sides by 12 x = \$0.50 per pouch
2	B	Find the unit rate for one gallon of gas. $\dfrac{128}{4} = \dfrac{x}{1}$ 128*1=4*x 128 = 4x Divide both sides by 4 x = 32 miles per gallon
3	D	Find the unit rate for one mile. $\dfrac{39}{6} = \dfrac{x}{1}$ 39*1=6*x 39 = 6x Divide both sides by 6 x = 6.5 or 6 minutes 30 seconds
4	A	Find the unit rate for one day. $\dfrac{114}{3} = \dfrac{x}{1}$ 114*1=3*x 114 = 3x Divide both sides by 3 x = 38 laps per day
5	C	The cost of the seven items plus \$15.00 should equal \$65.00. If the average cost per item is x, then 7x is the cost of all seven items. Therefore 7x + \$15.00 = \$65.00 can be used to find x.
6	B	Find the unit rate for one day. Geoff shot 150 arrows (\$15*10) $\dfrac{150}{5}$ days $= \dfrac{x}{1}$ 150*1=5*x 150 = 5x Divide both sides by 5 x = 30 arrows per day

Question No.	Answer	Detailed Explanations
7	C	Find the total number of cookies and divide by 7. The total number of cookies is $74 + 3 = 77$. The number of batches is 7. So the total number of cookies per batch can be found using the expression $(74 + 3)/7$.
8	B	Find the unit rate for one paper. Change hours to minutes or 3 hrs = 3*60 mins $$\frac{180}{124} = \frac{x}{1}$$ $180*1=124*x$ $180 = 124x$ Divide both sides by 124 $x = 1.45$ minutes or 1 minutes and 27 seconds
9	C	Find the unit rate for both and compare. Fink's: 462 miles/21 gallons = 22 miles per gallon Conrad's: 456 miles/12 gallons = 38 gallons Difference: $38 - 22 = 16$ gallons
10	D	Find the unit rate for both and compare. Myka: \$4.00/30 cards = \$0.13 per card Chuck: \$12.00/100 cards = \$0.12 per card Chuck paid 12 cents per card and Myka paid 13 cents per card. So Chuck got the better deal.
11	5	Since 3 cheeseburgers cost \$15, when you divide \$15 by 3, you get cost of one cheeseburger, which is \$5.

12

Day	Total Money Earned \$	Yards Mowed
Monday	50	2
Wednesday	75	3
Friday	25	1
Saturday	125	5

Tommy mowed 2 yards and made \$50, you divide the money earned by the yards mowed, 50/2 = \$ 25 per yard. Multiply each yard by \$25 to calculate how much money he earned each day.

13

	\$150	\$10	\$100	\$200
10			✓	
15	✓			
20				✓
1		✓		

10 tickets sold = \$100. 15 tickets sold = \$150. 20 tickets sold = \$200. 1 ticket sold = \$10.

Lesson 5: Finding Percent

Question No.	Answer	Detailed Explanations
1	B	$\dfrac{x}{24} = \dfrac{25}{100}$ x*100 = 24*25 100x = 600 Divide both sides by 100 x = 6
2	A	$\dfrac{x}{60} = \dfrac{15}{100}$ x*100 = 60*15 100x = 900 Divide both sides by 100 x = 9
3	C	$\dfrac{9}{72} = \dfrac{x}{100}$ 9*100 = 72*x 900 = 72x Divide both sides by 72 x = 12.5%
4	B	$\dfrac{x}{190} = \dfrac{30}{100}$ x*100 = 190*30 100x = 5,700 Divide both sides by 100 x = 57
5	C	$\dfrac{x}{280} = \dfrac{15}{100}$ x*100 = 280*15 100x = 4,200 Divide both sides by 100 x = 42 cards
6	A	$\dfrac{8}{x} = \dfrac{80}{100}$ 8*100 = x*80 800 = 80x Divide both sides by 80 x = 10 gallons

Question No.	Answer	Detailed Explanations
7	B	156 of 178 legos in percentage would be $\frac{156}{678}$ x 100 = 23%
8	D	8 out of 15 balls were hit into the outfield. The fraction of balls hit into the outfield is $\frac{8}{15}$. Therefore, converting into percentages, we get, $\frac{8}{15} \times 100 = x$ or $\frac{15}{8} = \frac{100}{x}$ Hence, the correct answer is (D)
9	B	32% of the 78 flowers are roses. Therefore, number of roses $= \frac{32}{100}$ x 78 = 24.96 which is approximately 25.
10	C	Total number of sheets used = 6 + 8 + 10 = 24 which is 30% of the total sheets (x). $\frac{30}{100}$ * X = 24 x = 80. The number of sheet not taken out is 80−24 = 56.

Question No. 11

Item Purchased	Original Price	Amount of Discount	Amount Paid
Video Game	$80	20%	**$64**
Movie Ticket	$14	**20%**	$11.20
Laptop	$1,000	**25%**	$750
Shoes	$55.00	10%	$49.5

Amount paid for video game = $64, Because $80 x 0.80 = $64 (20% discount means, one has to pay 80% of the original price. 80% = 0.80)

(2) Original price of the movie ticket = $14
Amount paid = $ 11.2
Discount = 14 - 11.2 = 2.8
% Discount /100 = Discount / original price
% Discount = 100 x (Discount / original price) = 100 x $(\frac{2.8}{14})$ = $\frac{280}{14}$ = 20%

(3) Discount for the Laptop = 1000 - 750 = 250
% Discount = 100 x (Discount / original price) = 100 x $(\frac{250}{1000})$ = $\frac{25000}{1000}$ = 25%

| 12 | A & D | The correct answer options are A and D.
 Option A. $90 x .25 = $22.50
 Option D. $560 x .25 = $140.00 |

Lesson 6: Measurement Conversion

Question No.	Answer	Detailed Explanations
1	B	There are 12 inches in a foot. 69 inches * $(\frac{1\ foot}{12\ inches})$ = $\frac{69}{12}$ = 5.75 feet
2	D	There are 4 quarts to a gallon. 7*4 = 28 quarts 28 + 3 = 31 quarts
3	D	There are 100 cm in a meter and 1000 meters in a kilometer. 3.7 km * $(\frac{1000\ m}{1\ km})$ * $(\frac{100\ cm}{1\ m})$ = 370,000 cm
4	B	There are 16 ounces per pound. 136 ounces * $(\frac{1\ lb}{16\ oz})$ = 8.5 pounds
5	C	There are 8 ounces per cup, 2 cups per pint, 2 pints per quart and 4 quarts per gallon. 5 gal * $(\frac{4\ qts}{1\ gal})$ * $(\frac{2\ pints}{1\ qt})$ * $(\frac{2\ cups}{1\ pt})$ * $(\frac{8\ oz}{1\ cup})$ = 640 ounces
6	B	Find the total length of the race in meters: 1284 + 1635 + 1473 = 4392 meters There are 1000 meters in 1 kilometer. 4392 m * $(\frac{1\ km}{1000\ m})$ = 4.392 = 4.4 km
7	C	Find the total number of minutes for the month: 1 h 3 m + 1 h 18 m + 55 m + 68 m = 63 m + 78 m + 55 m + 68 m = 264 minutes. There are 60 minutes in 1 hour. 264 min * $(\frac{1\ hr}{60\ min})$ = 4.4 hours
8	C	There are 60 minutes in 1 hour. 3 miles/18 minutes = x miles/60 minutes 3*60 = 18*x 180 = 18x Divide both side by 18 x = 10 miles
9	B	Find the perimeter by adding all four sides of the garden: 67 + 67 + 92 + 92 = 318 in There are 12 inches in a foot. 318 in * $(\frac{1\ foot}{12\ in})$ = 26.5 feet

Question No.	Answer	Detailed Explanations
10	B	There are 100 cm in a meter and 2.54 cm in 1 inch. 1.27 meters * $(\frac{100\ cm}{1\ m})$ * $(\frac{1\ in}{2.54\ cm})$ = 50 inches 50 in / 12 in = 4.17 feet = 4 feet 2 inches
11	B	8 fl oz = 1 cup
12	C	1000 meters/1 kilometer = x meters/16 kilometers 1000*16 = 1*x x = 16,000

13

1	3 L	**3000 ml**	Because 3 x 1000 =3000
2	**5 g**	5000 mg	Because 5000 ÷ 1000 = 5
3	**8 m**	8000 mm	Because 8000 ÷ 1000 = 8
4	12 L	**12,000 ml**	Because 12 x 1000 =12,000
5	20 g	**20,000 mg**	Because 20 x 1000=20,000

Chapter 2: The Number System

Lesson 1: Division of Fractions

1. **What is the quotient of 20 divided by one-fourth?**

   Ⓐ 80
   Ⓑ 24
   Ⓒ 5
   Ⓓ 15

2. **Calculate:** $1\frac{1}{2} \div \frac{3}{4} =$

   Ⓐ 4

   Ⓑ $\frac{1}{2}$

   Ⓒ $\frac{3}{4}$

   Ⓓ 2

3. **Calculate:** $3\frac{2}{3} \div 2\frac{1}{6} =$

   Ⓐ $\frac{8}{13}$

   Ⓑ $\frac{12}{13}$

   Ⓒ $1\frac{5}{13}$

   Ⓓ $1\frac{9}{13}$

4. Calculate: $2\dfrac{3}{4} \div \dfrac{11}{4} =$

    Ⓐ   1

    Ⓑ   2

    Ⓒ   3

    Ⓓ   4

5. Calculate: $\dfrac{7}{8} \div \dfrac{3}{4} =$

    Ⓐ $1\dfrac{1}{6}$

    Ⓑ   2

    Ⓒ $\dfrac{21}{32}$

    Ⓓ $\dfrac{5}{9}$

6. Calculate: $6\dfrac{3}{4} \div 1\dfrac{1}{8} =$

    Ⓐ $\dfrac{1}{6}$

    Ⓑ   4

    Ⓒ $5\dfrac{3}{4}$

    Ⓓ   6

7. **Complete the following division using mental math.**

    7 divided by $\dfrac{1}{5}$

    Ⓐ   35

    Ⓑ $\dfrac{7}{5}$

    Ⓒ $\dfrac{5}{7}$

    Ⓓ $\dfrac{1}{35}$

8. Complete the following division using mental math.

11 divided by $\frac{6}{6}$

Ⓐ $\frac{66}{66}$

Ⓑ $\frac{1}{11}$

Ⓒ 1

Ⓓ 11

9. What is the result when a fraction is multiplied by its reciprocal?

Ⓐ $\frac{1}{2}$

Ⓑ 10

Ⓒ 1

Ⓓ It cannot be determined.

10. Simplify the following problem. Do not solve.

$$\frac{14}{21} \div \frac{28}{7}$$

Ⓐ $\frac{14}{21} \div \frac{28}{7}$

Ⓑ $\frac{2}{3} \times \frac{1}{4}$

Ⓒ 1

Ⓓ 10

11. Which of the following is equal to $1 \div \frac{3}{4}$? Circle the correct answer choice.

Ⓐ $\frac{4}{3}$

Ⓑ $\frac{2}{4}$

Ⓒ $\frac{1}{3}$

12. Fill in the blank.

$\frac{1}{2} \div 4 =$ ___?

13. Which of the following is equal to $\frac{7}{2} \div \frac{2}{6}$? Circle the correct answer choice.

Ⓐ $\frac{9}{2}$

Ⓑ $\frac{5}{4}$

Ⓒ $\frac{42}{4}$

CHAPTER 2 → Lesson 2: Division of Whole Numbers

1. A team of 12 players got an award of $1,800 for winning a championship football game. If the captain of the team is allowed to keep $315, how much money would each of the other players get? (Assume they split it equally.)

   Ⓐ $135
   Ⓑ $125
   Ⓒ $150
   Ⓓ $123.75

2. Peter gets a salary of $125 per week. He wants to buy a new television that costs $3,960. If he saves $55 per week, which of the following expressions could he use to figure out how many weeks it will take him to save up enough money to buy the new TV?

   Ⓐ $3,960 ÷ ($125 − $55)
   Ⓑ $3,960 − ($125)($55)
   Ⓒ ($3,960 ÷ $125) ÷ $55
   Ⓓ $3,960 ÷ $55

3. An expert typist typed 9,000 words in two hours. How many words per minute did she type?

   Ⓐ 4,500 words per minute
   Ⓑ 150 words per minute
   Ⓒ 75 words per minute
   Ⓓ 38 words per minute

4. Bethany cut off 18 inches of her hair for "Locks of Love". (Locks of Love is a non profit organization that provides wigs to people who have lost their hair due to chemotherapy.) It took her 3 years to grow it back. How much did her hair grow each month?

   Ⓐ 1 inch
   Ⓑ 2 inches
   Ⓒ 0.25 inches
   Ⓓ 0.5 inches

5. On "Jeopardy," during the month of September, the champions won a total of $694,562. Assuming that there were 22 "Jeopardy" shows in September, what was the average amount won each day by the champions?

   Ⓐ $12,435
   Ⓑ $21,891
   Ⓒ $35,176
   Ⓓ $31,571

6. **A marching band wants to raise $20,000 at its annual fundraiser. If they sell tickets for $20 a piece, how many tickets will they have to sell?**

   Ⓐ 500
   Ⓑ 10,000
   Ⓒ 100
   Ⓓ 1,000

7. **A classroom needs 3,200 paper clips for a project. If there are 200 paper clips in a package, how many packages will they need in all?**

   Ⓐ 160
   Ⓑ 1,600
   Ⓒ 18
   Ⓓ 16

8. **A homebuilder is putting new shelves in each closet he is building. He has 2,592 shelves in his inventory. If each closet needs 108 shelves, how many closets can he build?**

   Ⓐ 2.4
   Ⓑ 108
   Ⓒ 42
   Ⓓ 24

9. **A toy maker needs to make $17,235 per month to meet his costs. Each toy sells for $45. How many toys does he need to sell in order to break even (cover his costs)?**

   Ⓐ 393
   Ⓑ 473
   Ⓒ 373
   Ⓓ 383

10. **A stamp collector collected 4,224 stamps last year. He collected the same amount each month. How many stamps did he collect each month?**

    Ⓐ 422
    Ⓑ 352
    Ⓒ 362
    Ⓓ 252

11. **Fill in the blank**

    40,950 ÷ _____ = 26

12. **Fill in the blank**

    455 ÷ _____ = 7

CHAPTER 2 → Lesson 3: Operations with Decimals

1. Three friends went out to lunch together. Ben got a meal that cost $7.25, Frank got a meal that cost $8.16, and Herman got a meal that cost $5.44. If they split the check evenly, how much did they each pay for lunch? (Assume no tax)

   Ⓐ $6.95
   Ⓑ $7.75
   Ⓒ $7.15
   Ⓓ $6.55

2. Which of these is the standard form of twenty and sixty-three thousandths?

   Ⓐ 20.63000
   Ⓑ 20.0063
   Ⓒ 20.63
   Ⓓ 20.063

3. Mr. Zito bought a bicycle for $160. He spent $12.50 on repair charges. If he sold the same bicycle for $215, what would his profit be on the investment?

   Ⓐ $ 147.50
   Ⓑ $ 42.50
   Ⓒ $ 67.50
   Ⓓ $ 55.00

4. A certain book is sold in a paperback version for $4.75 or in a hardcover version for $11.50. If a copy of the book is being purchased for each of the twenty students in Mrs. Jackson's class, how much money altogether would be saved by buying the paperback version, as opposed to the hardcover version?

   Ⓐ $ 155.00
   Ⓑ $ 135.00
   Ⓒ $ 115.00
   Ⓓ $ 145.00

5. **Which of these sets contains all equivalent numbers?**

Ⓐ $\left\{ 0.75, \dfrac{3}{4}, 75\%, \dfrac{8}{12} \right\}$

Ⓑ $\left\{ 0.100, \dfrac{5}{50}, 15\%, 0.010 \right\}$

Ⓒ $\left\{ \dfrac{3}{8}, 35\%, 0.35, \dfrac{35}{100} \right\}$

Ⓓ $\left\{ \dfrac{9}{25}, 36\%, 0.360, \dfrac{18}{50} \right\}$

6. **Brian is mowing his lawn. He and his family have 7.84 acres. Brian mows 1.29 acres on Monday, 0.85 acres on Tuesday, and 3.63 acres on Thursday. How many acres does Brian have left to mow?**

Ⓐ 2.70
Ⓑ 20.7
Ⓒ 2.07
Ⓓ 0.207

7. **Hector is planting his garden. He makes it 5.8 feet wide and 17.2 feet long. What is the area of Hector's garden?**

Ⓐ 9.976 square feet
Ⓑ 99.76 square feet
Ⓒ 99.76 feet
Ⓓ 997.6 square feet

8. **Chris and 2 of his friends go apple picking. Together they pick a bushel of apples that weighs 28.2 pounds. If Chris and his friends split the bushel of apples evenly among themselves, how many pounds of apples will each person take home?**

Ⓐ 9.4 pounds
Ⓑ 0.94 pounds
Ⓒ 94 pounds
Ⓓ 0.094 pounds

9. Joann and John are hiking over a three-day weekend. They have a total of 67.8 miles that they are planning on hiking. On Friday, they hike half of the miles. On Saturday, they hike another 20 miles. How many miles do they have left to hike on Sunday?

    Ⓐ   1.39 miles
    Ⓑ   31.9 miles
    Ⓒ   13.9 miles
    Ⓓ   139 miles

10. Margaret, Justin, and Leigh are babysitting for the neighbor's children during the summer. Each week they make a total of $72.00, and they split the money evenly. At the end of 4 weeks, how much money did Justin make?

    Ⓐ $130.00
    Ⓑ $144.00
    Ⓒ $72.00
    Ⓓ $96.00

11. Match the equation with the correct answer.

	18.711	1871.1	187.11
62.37 x 30 =			
6.237 x 30 =			
0.6237 x 30 =			

12. Fill in the blank.

$7.1 \times 3.2 =$ _____

13. Match the equation with the correct answer.

	6.67	88.84	910.54
9.13 – 2.46 =			
913 – 2.46 =			
91.3 – 2.46 =			

CHAPTER 2 → Lesson 4: Using Common Factors

1. **Which of these statements is true of the number 17?**

   Ⓐ It is a factor of 17.
   Ⓑ It is a multiple of 17.
   Ⓒ It is prime.
   Ⓓ All of the above are true.

2. **What are the single digit prime numbers?**

   Ⓐ 2, 3, 5, and 7
   Ⓑ 1, 2, 3, 5, and 7
   Ⓒ 3, 5, and 7
   Ⓓ 1, 3, 5, and 7

3. **Which of the following sets below contains only prime numbers?**

   Ⓐ 7, 11, 49
   Ⓑ 7, 37, 51
   Ⓒ 7, 23, 47
   Ⓓ 2, 29, 93

4. **The product of three numbers is equal to 105. If the first two numbers are 7 and 5, what is the third number?**

   Ⓐ 35
   Ⓑ 7
   Ⓒ 5
   Ⓓ 3

5. **What is the prime factorization of 240?**

   Ⓐ 2 × 2 × 2 × 5
   Ⓑ 2 × 2 × 2 × 2 × 15
   Ⓒ 2 × 2 × 2 × 2 × 5 × 3
   Ⓓ 2 × 2 × 2 × 5 × 3

6. **Office A's building complex and the school building next door have the same number of rooms. Office A's building complex has floors with 5 one-room offices on each, and the school building has 11 classrooms on each floor. What is the fewest number of rooms that each building can have?**

   Ⓐ 16
   Ⓑ 6
   Ⓒ 55
   Ⓓ $\frac{5}{11}$

7. How do you know if a number is divisible by 3?

Ⓐ if the ones digit is an even number
Ⓑ if the ones digit is 0 or 5
Ⓒ if the sum of the digits in the number is divisible by 3 or a multiple of 3
Ⓓ if the sum of the digits in the number is divisible by 2 and 3

8. Find the Greatest Common Factor (GCF) for 42 and 56.

Ⓐ 21
Ⓑ 14
Ⓒ 7
Ⓓ 2

9. What is the prime factorization of 110?

Ⓐ 10 × 11
Ⓑ 110 × 1
Ⓒ 55 × 2
Ⓓ 2 × 5 × 11

10. What is the LCM of 16 and 24?

Ⓐ 16
Ⓑ 24
Ⓒ 36
Ⓓ 48

11. Fill in the blank.

The Greatest Common Factor (GCF) of 24, 36, and 48 is _____.

12. List all the factors of 20.

13. Circle the common factors between 10 and 15.

    Ⓐ 1, 5

    Ⓑ 1, 2, 5

    Ⓒ 1, 3, 4

14. What are the common factors of 8 and 12? Write your answer in the box below.

CHAPTER 2 → Lesson 5: Positive and Negative Numbers

1. **Larissa has 4 $\frac{1}{2}$ cups of flour. She is making cookies using a recipe that calls for 2 $\frac{3}{4}$ cups of flour. After baking the cookies how much flour will be left?**

   Ⓐ 2 $\frac{3}{4}$ cups

   Ⓑ 2 $\frac{1}{4}$ cups

   Ⓒ 2 $\frac{3}{8}$ cups

   Ⓓ 1 $\frac{3}{4}$ cups

2. **The accounting ledger for the high school band showed a balance of $2,123. They purchased new uniforms for a total of $2,400. How much must they deposit into their account in order to prevent it from being overdrawn?**

   Ⓐ $382
   Ⓑ $462
   Ⓒ $4,000
   Ⓓ $277

3. **Juan is climbing a ladder. He begins on the first rung, climbs up four rungs, but then slides down two rungs. What rung is Juan on?**

   Ⓐ 2
   Ⓑ 3
   Ⓒ 4
   Ⓓ 5

4. **If last year Julie's net profit was $26,247 after she spent $14,256 on expenses, what was her gross revenue?**

   Ⓐ −$40,503
   Ⓑ −$11,991
   Ⓒ $40, 503
   Ⓓ $11,991

5. The amount of snow on the ground increased by 4 inches between 4 p.m. and 6 p.m. If there was 6 inches of snow on the ground at 4 p.m. how many inches were on the ground at 6 p.m.?

     Ⓐ 10 inches
     Ⓑ 14 inches
     Ⓒ 2 inches
     Ⓓ 18 inches

6. The temperature at noon was 20 degrees F. For the next 5 hours it dropped 2 degrees per hour. What was the temperature at 5:00 p.m.?

     Ⓐ 15 degrees F
     Ⓑ 10 degrees F
     Ⓒ 5 degrees F
     Ⓓ 0 degrees F

7. Tom enters an elevator that is in the basement, one floor below ground level. He travels three floors down to the parking level and then 4 floors back up. What floor does he end up on?

     Ⓐ ground level
     Ⓑ 2nd floor
     Ⓒ 3rd floor
     Ⓓ basement

8. Which of these numbers would not be found between 6 and 7 on a number line?

     Ⓐ $\dfrac{43}{6}$

     Ⓑ $\dfrac{34}{5}$

     Ⓒ $\dfrac{19}{3}$

     Ⓓ $\dfrac{100}{16}$

9. Stacey lives on a cliff. She lives 652 feet above sea level. When she travels to town, she travels down 491 feet. What elevation is town?

   Ⓐ 261 feet above sea level
   Ⓑ 161 feet below sea level
   Ⓒ 161 feet above sea level
   Ⓓ 141 feet above sea level

10. Chris lives in Alaska. When he wakes up, the temperature is −53 degrees F. By noon, the temperature has risen 27 degrees F. What is the temperature at noon?

    Ⓐ −27 degrees F
    Ⓑ −26 degrees F
    Ⓒ 26 degrees F
    Ⓓ −24 degrees F

11. Which of the following would result in a negative answer? Choose all that apply.

    Ⓐ 25 - 56
    Ⓑ 15 + 42
    Ⓒ -15 + 8
    Ⓓ -25 – 7
    Ⓔ 74 – 32

12. On a number line, how far apart are -27 and 30? Write your answer in the box below.

13. Which among the following has the lowest value?

    -15, 0, 10, 25

    Enter your answer in the box below.

CHAPTER 2 → Lesson 6: Representing Negative Numbers

1. **Which of these numbers would be found closest to 0 on a number line?**

    Ⓐ −5

    Ⓑ −5 $\frac{1}{2}$

    Ⓒ 4 $\frac{1}{2}$

    Ⓓ −4

2. **On a number line, how far apart are the numbers −5.5 and 7.5?**

    Ⓐ 13 units
    Ⓑ 12 units
    Ⓒ 12.5 units
    Ⓓ 2 units

3. **Which numbers does the following number line represent?**

    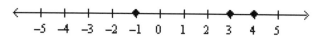

    Ⓐ {−2, 0, 5}
    Ⓑ {−3, −1, 4}
    Ⓒ {−1, 3, 5}
    Ⓓ {−1, 3, 4}

4. **If you go left on a number line, you will** _____

    Ⓐ go in a negative direction
    Ⓑ go in a positive direction
    Ⓒ increase your value
    Ⓓ none of these

5. **Which of the following numbers is NOT between −7 and −4 on the number line?**

    Ⓐ −7.2
    Ⓑ −4.8
    Ⓒ −6
    Ⓓ −5.01

6. **−7.25 is between which two numbers on the number line?**

   Ⓐ −7 and −6
   Ⓑ −5 and −3
   Ⓒ −9 and −10
   Ⓓ −7 and −8

7. **What happens when you start at any number on a number line and add its additive inverse?**

   Ⓐ The number doubles.
   Ⓑ The number halves.
   Ⓒ The sum is zero.
   Ⓓ There is no movement.

8. **Which set of numbers would be found to the left of 4 on the number line?**

   Ⓐ {−1, 4, −5}
   Ⓑ {1, −4, −5}
   Ⓒ {1, 4, −5}
   Ⓓ {1, 4, 5}

9. **On a number line, Bill places a marble on 12. He rolls the marble to the left and it moves 21 spaces and then rolls back to the right 3 spaces. What number does the marble end up on?**

   Ⓐ −21
   Ⓑ −8
   Ⓒ −9
   Ⓓ −6

10. **Which number would be found on the number line between 0 and −1?**

    Ⓐ −1.2
    Ⓑ 1.2
    Ⓒ −0.8
    Ⓓ 0.8

11. **Circle the number with the highest value.**

    Ⓐ −17
    Ⓑ −28
    Ⓒ −36

12. Which number(s) have a value lower than 3? Select all that apply.

    Ⓐ −5

    Ⓑ 6

    Ⓒ −18

    Ⓓ −23

    Ⓔ 4

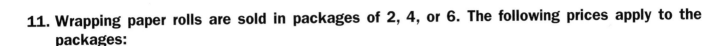

11. **Wrapping paper rolls are sold in packages of 2, 4, or 6. The following prices apply to the packages:**

Package of 2 rolls —— $1.50
Package of 4 rolls —— $2.50
Package of 6 rolls —— $3.50

Lisa needs to buy exactly 12 rolls of wrapping paper for a service project.
Below are a few of the possible combinations. Select the value that correctly represents the missing number in the combinations listed on the table.

	2	4	6
2, 2, 2, 2, 2, **a** or 6 ($1.50) = $9.00: a =	O	O	O
2, 2, 2, **b**, 4 or 4($1.50)) + $2.50 = $8.50: b =	O	O	O
2, 2, **c**, 4, or 2($1.50) + 2($2.50) = $8.00: c =	O	O	O

12. **The height of the four tallest downtown buildings of three different cities is given in the table below. Calculate the mean height of each city's buildings and match it with the correct answer from the bottom table.**

City A Buildings	Height (m)
1	23
2	32
3	35
4	45

City B Buildings	Height (m)
1	15
2	22
3	28
4	25

City C Buildings	Height (m)
1	15
2	30
3	35
4	36

	22.50 m	29 m	23.20 m	33.75 m	18 m
City A Buildings	O	O	O	O	O
City B Buildings	O	O	O	O	O
City C Buildings	O	O	O	O	O

CHAPTER 5 → Lesson 3: Central Tendency

1   Jason was conducting a scientific experiment using bean plants. He measured the height (in centimeters) of each plant after three weeks. These were his measurements (in cm): 12, 15, 11, 17, 19, 21, 13, 11, 16

What is the average (mean) height? What is the median height?

Ⓐ Mean = 15 cm, Median = 19 cm
Ⓑ Mean = 15 cm, Median = 15 cm
Ⓒ Mean = 19 cm, Median = 15 cm
Ⓓ Mean = 15 cm, Median = 13 cm

2. Stacy has 60 pairs of shoes. She has shoes that have a heel height of between 1 inch and 4 inches. Stacy has 20 pairs of shoes that have a 1 inch heel, 15 pairs of shoes that have a 2 inch heel and 20 pairs of shoes that have a 3 inch heel height. Remaining shoes have 4 inch heel height. What is the average heel height of all 60 pairs of Stacy's shoes?

Ⓐ 1.3 inches
Ⓑ 1.5 inches
Ⓒ 2 inches
Ⓓ 2.2 inches

3. What is the median of the following set of numbers?
{16, –10, 13, –8, –1, 5, 7, 10}
Ⓐ –8
Ⓑ –1
Ⓒ 4
Ⓓ 6

4. Given the following set of data, is the median or the mode larger?
{5, –10, 14, 6, 8, –2, 11, 3, 6}
Ⓐ The mode
Ⓑ The median
Ⓒ They are the same
Ⓓ You cannot figure it out

5. A = { 10, 15, 2, 14, 19, 25, 0 }
Which of the following numbers, if added to Set A, would have the greatest effect on its median?
Ⓐ 14
Ⓑ 50
Ⓒ 5
Ⓓ 15

6. **What is the mean, median and mode for the following data?**

   {−7, 18, 29, 4, −3, 11, 22}

   Ⓐ Mean = 10
   Median = 18
   Mode = −3

   Ⓑ Mean = 11
   Median = 11
   Mode = none

   Ⓒ Mean = 10.57
   Median = 11
   Mode = none

   Ⓓ Mean = 10
   Median = 18
   Mode = −3

7. **Amanda had the following numbers: 1,2,6**

   If she added the number 3 to the list...

   Ⓐ the mean would increase
   Ⓑ the mean would decrease
   Ⓒ the median would increase
   Ⓓ the median would decrease

8. **A pizza shop sells the following ice cream treats:**

   Strawberry Gelato: $0.60
   Chocolate Cone: $0.80
   Lemon Ice: $0.45

   What is the average sale price for the ice cream treats?

   Ⓐ $0.62
   Ⓑ $0.45
   Ⓒ $0.60
   Ⓓ $0.80

9. Marcel has 5 stamp collections. He wants to average 35 stamps per collection. So far, he has 28, 62, 12, and 44 stamps in each collection. How many stamps does he need to have in his 5th collection to average 35 stamps?

Ⓐ 30
Ⓑ 29
Ⓒ 28
Ⓓ 27

10. How much will the mean of the following set of numbers increase if the number 53 is added to it?
{67, 29, 40, –12, 88, –7, 11}

Ⓐ 53
Ⓑ 2.768
Ⓒ 2.571
Ⓓ 6.625

11. Find the mean of the following set.
{ 25.1, 19.6, 88.5, 0, -3, 19.8 }
Circle the correct answer.

Ⓐ 19.6
Ⓑ 26
Ⓒ 25
Ⓓ 19.7

12. Select the median number for each set of numbers.

	3	4	8
4, 2, 3, 6, 4, 9, 7	○	○	○
6, 3, 2, 8, 1, 3, 6	○	○	○
8, 9, 4, 8, 1, 10, 3	○	○	○

13. Select the median number for each set of numbers.

	11	15	18
12, 18, 11, 20, 20	○	○	○
14, 20, 18, 11, 15	○	○	○
8, 6, 15, 11, 18	○	○	○

CHAPTER 5 → Lesson 4: Graphs & Charts

1. **Which of the following graphs best represents the values in this table?**

X	Y
1	1
2	3
3	1
4	3

Ⓐ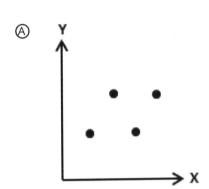

Ⓒ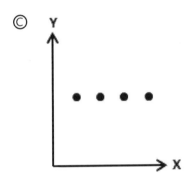

Ⓑ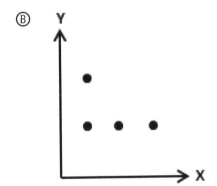

Ⓓ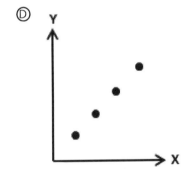

2. The results of the class' most recent science test are displayed in this histogram. Use the results to answer the question. A "passing" score is 61 or higher.

How many students passed the science test?

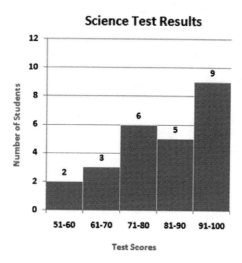

Ⓐ 3 students
Ⓑ 20 students
Ⓒ 23 students
Ⓓ 25 students

3. The results of the class' most recent science test are displayed in this histogram. Use the results to answer the question. How many students scored a 90 or below?

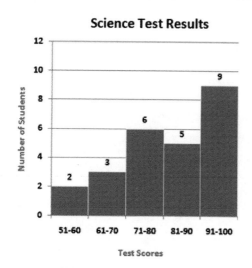

Ⓐ 5 students
Ⓑ 9 students
Ⓒ 11 students
Ⓓ 16 students

4. As part of their weather unit, the students in Mr. Green's class prepared a line graph showing the high and low temperatures recorded each day during a one-week period. Use the graph to answer the question.

On which day was the greatest range in temperature seen?

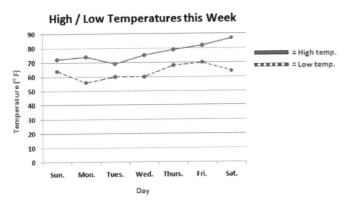

Ⓐ Sunday
Ⓑ Monday
Ⓒ Friday
Ⓓ Saturday

5. The sixth graders at Kilmer Middle School can choose to participate in one of the four music activities offered. The number of students participating in each activity is shown in the bar graph below. Use the information shown to answer the question.

There are 180 sixth graders in the school. About how many do not participate in one of the music activities?

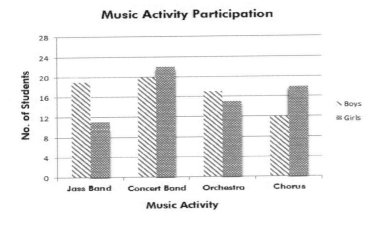

Ⓐ 66 students
Ⓑ 46 students
Ⓒ 25 students
Ⓓ 15 students

6. **A.J. has downloaded 400 songs onto his computer. The songs are from a variety of genres. The circle graph below shows the breakdown of his collection by genre. Use the information shown to answer the question.**

Which two genres together make up more than half of A.J.'s collection?

A.J.'s Music Collections

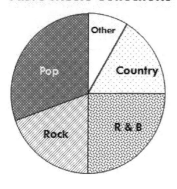

Ⓐ R + B and Country
Ⓑ Country and Rock
Ⓒ Rock and Pop
Ⓓ Pop and R + B

7. **The results of the class' most recent science test are displayed in this histogram. Use the results to answer the question.**

What percentage of the class scored an 81-90 on the test?

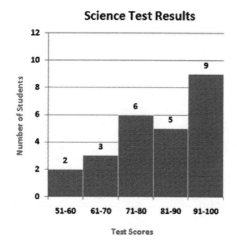

Ⓐ 5%
Ⓑ 20%
Ⓒ 25%
Ⓓ 30%

8. How much of the graph do undergarments and socks make up together?

Clothing Sales Breakdown

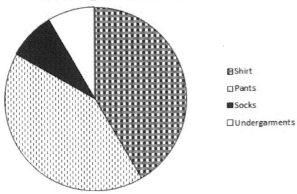

- ⊞ Shirt
- ☐ Pants
- ■ Socks
- ☐ Undergarments

Ⓐ less than 5%
Ⓑ between 5% and 10%
Ⓒ between 10% and 25%
Ⓓ more than 25%

9. As part of their weather unit, the students in Mr. Green's class prepared a line graph showing the high and low temperatures recorded each day during a one-week period. Use the graph to answer the question.

What percentage of the days had a high temperature of 80 degrees or higher? Round to the nearest tenth.

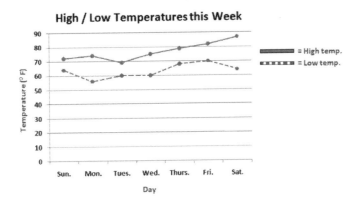

Ⓐ 14.3%
Ⓑ 42.9%
Ⓒ 28.6%
Ⓓ None of these

10. The sixth graders at Kilmer Middle School can choose to participate in one of the four music activities offered. The number of students participating in each activity is shown in the bar graph below. Use the information shown to answer the question.

There are 132 students who participate in music activities. What percentage of students who participate in music activities participate in chorus or jazz band?

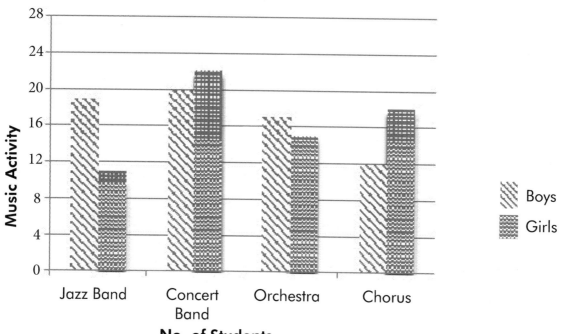

Ⓐ 45%
Ⓑ 38%
Ⓒ 40%
Ⓓ 50%

Questions 11 and 12 are based on the bar graph shown below.

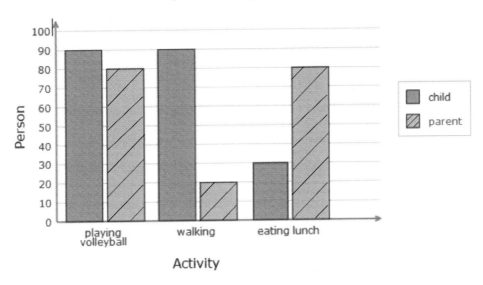

Family beach day

11. **Which of the following statements are true? Select all that apply.**

   Ⓐ The same number of children participated in playing volleyball and walking
   Ⓑ The same number of children participated in walking and eating lunch.
   Ⓒ All of the children who played volleyball also ate lunch.
   Ⓓ More children than parents ate lunch.
   Ⓔ More children than parents played volleyball.
   Ⓕ The same number of parents played volleyball as ate lunch.

12. **What is the total number of parents and children who played volleyball and ate lunch? Write your answer in the box below.**

CHAPTER 5 → Lesson 5: Data Interpretation

1. In the last 4 seasons, Luis scored 14, 18, 15, and 25 goals. How many goals does he need to score this year, to end the season with an overall average of 20 goals per season?

   Ⓐ 29
   Ⓑ 27
   Ⓒ 30
   Ⓓ 28

2. Stacy has 60 pairs of shoes. She has shoes that have a heel height of between 1 inch and 4 inches. Stacy wants to know which heel height she has the most of. Would she figure out the mode or mean?

   Ⓐ Mean
   Ⓑ Mode
   Ⓒ Both
   Ⓓ Neither

3. There are three ice cream stands within 15 miles and they are owned by Mr. Sno.

Ice Cream Stand	Vanilla	Chocolate	Twist
A	15	22	10
B	24	8	14
C	20	16	13

   What percentage of the ice cream cones sold by Ice Cream Stand B were vanilla? Round your answer to the nearest whole number.

   Ⓐ 52%
   Ⓑ 24%
   Ⓒ 50%
   Ⓓ 25%

4. There are three ice cream stands within 15 miles and they are owned by Mr. Sno.

Ice Cream Stand	Vanilla	Chocolate	Twist
A	15	22	10
B	24	8	14
C	20	16	13

   What percentage of the ice cream cones sold by Ice Cream Stand C were chocolate? Round your answer to the nearest whole number.

   Ⓐ 30%
   Ⓑ 32%
   Ⓒ 33%
   Ⓓ 62%

5. Alexander plays baseball. His batting averages for the games he played this year were recorded. What is the batting average he had the most often?

{.228, .316, .225, .333, .228, .125, .750, .500}

Ⓐ .228
Ⓑ .316
Ⓒ .750
Ⓓ .333

6. How much will the mean increase by when the number 17 is added to the set? Round your numbers to the nearest tenth.

{5, -10, 14, 6, 8, -2, 11, 3, 6}

Ⓐ 1.5
Ⓑ 1
Ⓒ 1.2
Ⓓ 1.7

7. Andrea plays the violin. She is practicing songs for her concert and she times how long it takes her to play each song. She wants to put the song with the middle number of minutes at the beginning of her concert. How many minutes is the song that she will play first?

{13, 8, 4, 16, 3, 9, 11}

Ⓐ 13 minutes
Ⓑ 11 minutes
Ⓒ 8 minutes
Ⓓ 9 minutes

8. There are three ice cream stands within 15 miles and they are owned by Mr. Sno.

Ice Cream Stand	Vanilla	Chocolate	Twist
A	15	22	10
B	20	8	14
C	24	16	13

What percentage of the total ice cream cones sold were twist? Round your answer to the nearest whole number.

Ⓐ 26%
Ⓑ 24%
Ⓒ 62%
Ⓓ 14%

9. Amy is trying to find out the number of magazines that have a woman on the cover. She goes to the book store and looks at the rack of magazines. There are at least 30 magazines. Amy looks at the covers of 20 magazines. Did Amy get a good sample?

Ⓐ Yes, because she looked at all of the magazines.
Ⓑ Yes, because she got a sample of more than half of the magazines.
Ⓒ No, because she did not looked at all of the covers.
Ⓓ No, because she looked at less than half of the magazines.

10. Travis is putting together outfits for work. He has 4 red shirts, 8 blue shirts and 7 white shirts. He has 6 pairs of khaki pants. What percent of Travis' possible outfits have blue shirts? Round your answer to the nearest whole number.

Ⓐ 8%
Ⓑ 43%
Ⓒ 42%
Ⓓ 48%

11. Travis is putting together outfits for work. He has 4 red shirts, 8 blue shirts and 7 white shirts. He has 6 pairs of khaki pants. What percent of Travis' possible outfits do not have red shirts?

12. Scientists were concerned about the survival of the Mississippi Blue Catfish, so they collected data from samples of this species of fish. The scientists captured the fish, measured them, and then returned them to the lake from which they were taken. Select the correct numbers for different ranges of lengths.

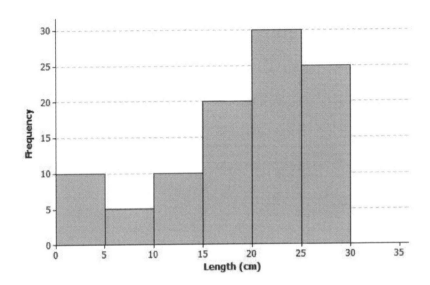

	5	10	20	25	30
0 – <5 cm	O	O	O	O	O
5 – <10 cm	O	O	O	O	O
10 – <15 cm	O	O	O	O	O
15 – <20 cm	O	O	O	O	O
20 – <25 cm	O	O	O	O	O

13. Each statement below describes a sample of data. Select all the statements that describe the situations that would result in non-biased or equal representation of the data.

Ⓐ Bella randomly called 100 phone numbers in her town to survey them about their shopping habits. 20 people chose to participate.

Ⓑ Heather climbed 11 trees throughout each forest in her county. Each forest is the same size.

Ⓒ Arujun bought 2 eggs from each farmer at the farmer's market. Each farmer had an equal number of eggs.

Ⓓ Allen put the names of all the cities in a state into a jar. Then he drew out 32 names from the jar.

Ⓔ Kiera polled 12 people at the town senior center.

CHAPTER 5 → Lesson 6: Describing the Nature

1. A new sandwich shop just opened. It is offering a choice of ham sandwiches, chicken sandwiches, or hamburgers as the main course and French fries, potato salad, baked beans or coleslaw to go with them. How many different ways can a customer have a sandwich and one side order?

   Ⓐ 6
   Ⓑ 8
   Ⓒ 10
   Ⓓ 12

2. Karen has a set of numbers that she is working with {6, 14, 28, 44, 2, −6}. What will happen to the mean if she adds the number −8 to the set?

   Ⓐ The mean will decrease.
   Ⓑ The mean will increase.
   Ⓒ The mean will stay the same.
   Ⓓ It cannot be determined.

3. Susan goes to the store to buy supplies to make a cake. What is the average amount that Susan spent on each of the ingredients she bought?
   {$4.89, $2.13, $1.10, $3.75, $0.98, $2.46}

   Ⓐ $5.22
   Ⓑ $2.55
   Ⓒ $2.25
   Ⓓ $2.50

4. Robert had an average of 87.0 on his nine math tests. His scores on the first eight tests were {92, 96, 83, 81, 94, 78, 93, 70}. What score did Robert receive on his last test?

   Ⓐ 90
   Ⓑ 89
   Ⓒ 96
   Ⓓ 87

5. Terry is playing a game with his brother and they played 5 times. The median score that Terry got was 37. The range of Terry's scores was 40. What was the lowest and highest score that Terry got?

   Ⓐ 37 and 77
   Ⓑ 10 and 50
   Ⓒ 17 and 57
   Ⓓ Not enough information is given.

6. Marcus wants to find out how many people go to the zoo on Saturdays in August. The zoo is open for 7 hours and there are 4 Saturdays in August. He counts the number of people who enter the zoo for two hours one Saturday. Will Marcus get an accurate idea of how many people go to the zoo?

Ⓐ Yes because he will get a good representative sample.
Ⓑ No because he will not get enough of a representative sample.
Ⓒ Yes because he can assume that two hours is enough time to count the number of people.
Ⓓ No because he would need to be there for less time.

7. Carl surveys his class to find out how tall his classmates are.
How many classmates did Carl survey?

Height	Number of Classmates
4'4" – 4'8"	1
4'9" – 5'	4
5'1" – 5'5"	7
5'6" – 5'11"	14
6' – 6'4"	2

Ⓐ 27
Ⓑ 28
Ⓒ 14
Ⓓ 24

8. Julie is making a quilt. She uses 7 different patterns for her quilt squares.

What percentage of Julie's quilt is not flowers or butterflies?

Squares	Number
Solid	15
Stripe	22
Flower	8
Plaid	12
Zig-Zag	13
Circles	10
Butterflies	6

Ⓐ 80%
Ⓑ 72%
Ⓒ 84%
Ⓓ 83%

9. Carl surveys his class to find out how tall his classmates are.

   How many classmates are taller than 5 feet?

Height	Number of Classmates
4'4" – 4'8"	1
4'9" – 5'	4
5'1" – 5'5"	7
5'6" – 5'11"	14
6' – 6'4"	2

   Ⓐ 23
   Ⓑ 14
   Ⓒ 27
   Ⓓ 16

10. Travis is putting together outfits for work. He has 4 red shirts, 8 blue shirts and 7 white shirts. He has 6 pairs of khaki pants. What percent of Travis' possible outfits do not have red shirts?

    Ⓐ 79%
    Ⓑ 90%
    Ⓒ 78%
    Ⓓ 15%

11. What is the mean of these numbers? 5, 4, 3, 7, 8, 9. Circle the correct answer choice.

    Ⓐ 6
    Ⓑ 5
    Ⓒ 7
    Ⓓ 3

12. Dana's recent math exams grades were: 67, 62, 70, 69, 90, 93, 95, 88, and 93. What is Dana's median score? Circle the answer that represents the median score.

    Ⓐ 88
    Ⓑ 93
    Ⓒ 67

CHAPTER 5 → Lesson 7: Context of Data Gathered

1. Bella recorded the number of newspapers left over at the end of each day for a week. What is the average number of newspapers left each day?

    3, 6, 1, 0, 2, 3, 6

    Ⓐ 2 newspapers
    Ⓑ 3 newspapers
    Ⓒ 4.3 newspapers
    Ⓓ 6 newspapers

2. Recorded below are the ages of eight friends. Which statement is true about this data set?

    13, 18, 14, 12, 15, 19, 11, 15

    Ⓐ Mean > Median > Mode
    Ⓑ Mode < Mean < Median
    Ⓒ Median > Mode < Median
    Ⓓ Mode > Mean > Median

3. The line plot below represents the high temperature in °F each day of a rafting trip. What was the average high temperature during the trip?

    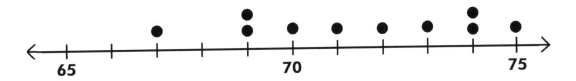

    Ⓐ 70°F
    Ⓑ 71°F
    Ⓒ 71.4°F
    Ⓓ 72°F

4. What is the mean absolute deviation of the grades Charlene received on her first ten quizzes: 83%, 92%, 76%, 87%, 89%, 96%, 88%, 91%, 79%, 99%.

Ⓐ 4.5%
Ⓑ 5.4%
Ⓒ 88%
Ⓓ 88.5%

5. The following numbers are the minutes it took Nyak to complete one lap around a certain dirt bike course. What is the mean absolute deviation of this data set?

12.3, 12.6, 12.2, 10.9, 11.3, 10.3, 11.7, 10.7

Ⓐ 0.7 minutes
Ⓑ 0.75 minutes
Ⓒ 11.1 minutes
Ⓓ 1.5 minutes

6. The local Akita Rescue Organization has nine Akitas for adoption. Their weights are 82 lb, 95 lb, 130 lb, 112 lb, 122 lb, 72 lb, 86 lb, 145 lb, 93 lb. What are the median (M), 1st Quartile (Q1), 3rd Quartile (Q3) and Interquartile Range (IQR) of this data set?

Ⓐ M = 95, Q1 = 84, Q3 = 126, IQR = 42
Ⓑ M = 95, Q1 = 85, Q3 = 126, IQR = 145
Ⓒ M = 95, Q1 = 84, Q3 = 145, IQR = 42
Ⓓ M = 95, Q1 = 126, Q3 = 145, IQR = 42

7. Monica is keeping track of how much money she spends on lunch each school day. This week Monica spent $5.20, $6.50, $3.75, $0.75, and $4.15. What is the median (M) and Interquartile Range (IQR) of Monica's lunch expense?

Ⓐ M = $4.15, IQR = $1.45
Ⓑ M = $4.07, IQR = $1.45
Ⓒ M = $4.07, IQR = $2.25
Ⓓ M = $4.15, IQR = $3.60

8. Monica's friend Anna is also tracking her school lunch expenses. This week Anna spent $2.60, $0, $7.00, $4.40, $3.75. What is the difference between Monica's and Anna's average daily lunch expense?

Monica: $5.20, $6.50, $3.75, $0.75, and $4.15

Ⓐ $0.40
Ⓑ $0.52
Ⓒ $0.60
Ⓓ $1.50

9. **Zahra and Tyland are very competitive. Each have recorded their basketball scores since the beginning of the season. Which data set has a greater Interquartile Range (IQR)?**

Name	Scores
Zahra	8, 10, 3, 12, 14, 11
Tyland	6, 7, 18, 12, 4, 9

Ⓐ Zahra has the higher IQR of 4.
Ⓑ Zahra has the higher IQR of 10.5.
Ⓒ Tyland's IQR is 2 points higher than Zahra's.
Ⓓ Tyland and Zahra have the same IQR of 5.

10. **Jamila records the number of customers she gets each day for a week: 20, 42, 15, 23, 53, 62, 58. What is the mean absolute deviation of this data?**

Ⓐ 16.9 customers
Ⓑ 19 customers
Ⓒ 39 customers
Ⓓ 47.5 customers

11. **Frank played 9 basketball games this season. His scores were 15, 20, 14, 36, 20, 10, 35, 23, and 24. What is the median of Frank's scores? Write the answer in the box.**

12. **Gordan has the following data: 13, 19, 16, z, 17**
If the mean is 16, which number could be z? Circle the correct answer.

Ⓐ 11
Ⓑ 15
Ⓒ 18

CHAPTER 5 → Lesson 8: Relating Data Distributions

1. **The histogram below shows the grades in Mr. Didonato's history class.**

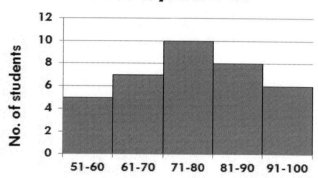

    Which of the following statements is true based on this data?

    Ⓐ The mode score is 75%.
    Ⓑ More than half the students received 81% or higher.
    Ⓒ Ten students received a 70% or lower.
    Ⓓ There is not enough information to determine the median score.

2. **The bar graph below shows the different types of shoes sold by *Active Feet*.**

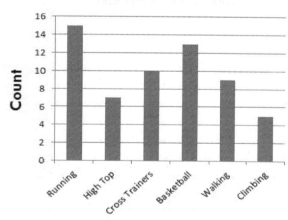

    Which of the following statements is true based on this data?

    Ⓐ The store carries more types of basketball and climbing shoes combined than types of running shoes.
    Ⓑ The store sells more running shoes than any other shoe type.
    Ⓒ The store has more types of walking shoes than cross trainers.
    Ⓓ The store earns more money selling basketball shoes than walking shoes.

3. Milo and Jacque went fishing and recorded the weight of their fish in the line plots below.

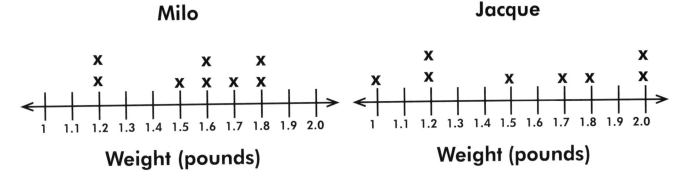

Milo

Weight (pounds)

Jacque

Weight (pounds)

Which of the following statements is true based on this data?

Ⓐ The weights of Milo's fish have more variability.
Ⓑ The average weight of Jacque's fish is greater than the average weight of Milo's fish.
Ⓒ The weights of Jacque's fish have a greater mean absolute deviation.
Ⓓ Milo caught more fish by weight than Jacque.

4. The frequency table below records the age of the student who attended the dance.

Age	Frequency
11	17
12	20
13	9
14	8
15	12
16	4

Which of the following statements is true based on this data?

Ⓐ 80 students attended the dance.
Ⓑ The mean age is 12.86
Ⓒ If 3 more 16-year-olds arrive, the median age would increase.
Ⓓ More than half the students are older than the mean.

5. For twelve weeks Jeb has recorded his car's fuel mileage in miles per gallon (mpg). He has plotted these fuel mileages below in a box plot.

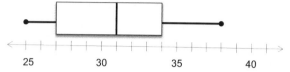

Which of the following statements is true based on this data?

Ⓐ The best fuel mileage recorded was 34 mpg.
Ⓑ Half of the data collected was between 27 mpg and 34 mpg.
Ⓒ The median fuel mileage was 30.5 mpg.
Ⓓ Three data points lie in the Interquartile range.

6. *"Forever Green"* recorded the heights of the pine trees sold this month in the box plot shown below. There are 15 data points in Quartile 3.

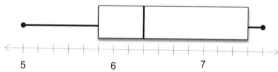

Which of the following statements is true based on this data?

Ⓐ About 25% of the trees were 7 feet 3 inches or taller.
Ⓑ There are more data points in Quartile 2 than Quartile 3.
Ⓒ The median height is 6 feet 2 inches.
Ⓓ "Forever Green" sold a total of about 60 trees this month.

7. As part of a math project, each student in Ms. Lauzon's class recorded the number of jumping jacks they could perform in 30 seconds.

Stem	Leaf
3	0 0 2 4 7 7
4	1 1 4 7 9
5	2 3 4 4 6 6 7
6	0 0 1 1 3 8 8
7	1 2
3|0 means 30	

Which of the following statements is true based on this data?

Ⓐ There are 21 students in Ms. Lauzon's class.
Ⓑ The median number of jumping jacks performed was 53.
Ⓒ The range of jumping jacks is 42.
Ⓓ Seven people performed less than 40 jumping jacks.

8. Tamryn and Clio both make beaded necklaces. The line plots below display the number of beads on ten necklaces each from Tamryn and Clio.

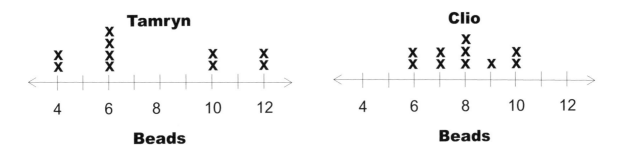

Which of the following statements is true based on this data?

Ⓐ If Tamryn added an eleventh data point of 10, the mean would not change.
Ⓑ On an average, Tamryn uses more beads than Clio.
Ⓒ Clio has more variability in the number of beads she uses.
Ⓓ If Clio added an eleventh data point of 8, the median would stay the same.

9. The histogram below shows the high temperature each day for a month.

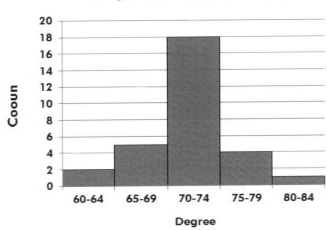

Which of the following statements is true based on this data?

Ⓐ More than half the days had a temperature of 70–74°.
Ⓑ 20% of the days had temperatures above 74°.
Ⓒ The average temperature was 72°.
Ⓓ There were fewer days below 70° than days above 74°.

10. Marcel enjoys catching fireflies. The steam-and-leaf plot shows the number of fireflies Marcel caught on several nights.

Stem	Leaf
0	4 6 9
1	0 3 4 7 7 7 8
2	0 0 2 2 5
3	1 4
1|3 means 13	

Which of the following statements is true based on this data?

Ⓐ The median is greater than the mean.
Ⓑ The median number of fireflies caught was 17.
Ⓒ The range of fireflies is 34.
Ⓓ Marcel caught fireflies on 14 nights.

11. The students in one math class were asked how many brothers and sisters (siblings) they each have. The plot below shows the distribution of the data (answers from the students). Circle the blue box that represents the number of siblings had by the most students in this class.

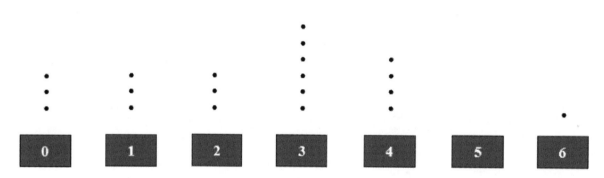

12. Three volleyball teams each recorded their scores for their first 5 games. Use each team's scores below to determine the missing value.

	10	14	17
Team 1: 15, 16, t, 17, 12, Mode = 17, What is t?	○	○	○
Team 2: 7, 14, r, 16, 13 Mean = 12, What is r?	○	○	○
Team 3: 11, 7, 19, 14, z, Median = 14, What is z?	○	○	○

End of Statistics & Probability

Chapter 5
Statistics & Probability

Lesson 1: Statistical Questions

Question No.	Answer	Detailed Explanations
1	B	Add the number of boys who participated in each activity to find the total number of boys. 19 + 20 + 17 + 12 = 68
2	C	When conducting a survey, it is most accurate to ask the question to the focus group you are trying to reach.
3	D	Taking the temperature consistently at the warmest and coolest time of the day will provide a consistent data sample for Emily's survey.
4	B	A representative sample should be a sample of people who represent the larger population. Surveying all families in one neighborhood would not be representative because people in the same neighborhood most likely have the same income level.
5	D	A bar graph would be best because it shows the data in bars that are then very easy to compare.
6	C	Derek's survey will be biased because the students in his class have the same amount of homework as he does. To get an unbiased sample, he should ask students from all fourth grade classes.
7	C	A bar graph will show the average snowfall of each city in a bar of different colors. Those bars will be very easy to compare and will make it easy to compare the average snowfall.
8	A	A circle graph is the best graph to use when representing percentages. The circle graph would show what percentage of a family's entire day is spent doing each activity.
9	A	The randomly selected cards are not biased. A biased sample is one that contains the same or very similar items. Randomly selected samples are generally not biased.

Question No.	Answer	Detailed Explanations
10	C	A line graph would be best because Cara can plot the miles she runs and connect them with line segments. She will then be able to visually see the fluctuation in the miles (the trend) in the distance she runs.
11	B, C, D & E	Options (B), (C), (D), and (E) are correct, A statistical question is one that can be answered by collecting data that vary (i.e., not all of the data values are the same).
12		

Question	Dot Plot
1. What are the ages of 4th graders in our school?	A
2. What are the heights of the players on the 8th grade boys' basketball team?	D
3. How many hours do 6th graders in our class watch TV on a school night?	C
4. How many different languages do students in our class speak?	B

1. On the average, ages of 4th graders vary between 9 to 12 (most of them are 10 years old). Therefore, we can say that Dot Plot A represents the ages of 4th graders.

2. Heights of the students have more variations. Therefore, we can say that Dot Plot D represents the heights of 8th grade boys' basketball team (and note also that in Dot Plot D, there are 12 points, which correspond to 12 players in a basketball team).

3. Number of hours 6th graders watch TV may vary between 1 to 3 hour, which is represented in Dot Plot C.

4. In a class, there may be students who speak different languages, which may vary from 1 to 6, which is represented in Dot Plot B.

(In all the Dot plots, note also how range of x- values are different for different quantities such as age, number of languages, hour, and height.)

Lesson 2: Distribution

Question No.	Answer	Detailed Explanations
1	B	The Socks section has a measure of 30 degrees and the Pants section has a measure of 145 degrees. Solve for the total pant sales by setting up a proportion and solving. $\dfrac{145}{30} = \dfrac{x}{\$60}$ $(145)(\$60) = 30x$ $8700 = 30x$ $290 = x$ Therefore, 300 is the best estimate.
2	C	The graph shows that there are 19 boys and 11 girls participating in the Jazz Band. $19 + 11 = 30$ sixth graders altogether
3	A	Angle corresponding to R + B songs = 90 degrees. The number of R + B songs downloaded is $400 \times (\dfrac{90}{360}) = 400 \times (\dfrac{1}{4}) = 100$ songs $(\dfrac{90}{360} = \dfrac{1}{4}$ after taking GCF 90 out of both the numerator and denominator) Angle correspoding to rock songs is close to 90 degrees but less than 90 degrees. Let us take it to be 75 degrees (approximately) So, number of rock songs downloaded is $400 \times (\dfrac{75}{360}) = 400 \times (\dfrac{5}{24}) = 83$ (approximately) $(\dfrac{75}{360} = \dfrac{5}{24}$ after taking GCF 15 out of both the numerator and denominator) Therefore A.J. has downloaded $100 - 83 = 17$ more R + B songs than rock songs. Among the choices given, (A) is the most appropriate choice.
4	A	To find the longest distance Colleen traveled, add the shortest distance to the range. $63 + 98 = 161$
5	B	To find the least number of letters that Bob has ever delivered, subtract the range from the largest number. $8,476 - 6,930 = 1,546$ letters

Question No.	Answer	Detailed Explanations
6	B	The median price is $280 and there are 66 pieces of art for sale. That means that half of the pieces of art have a price of $280 or less. Fred has $280 to spend. That means that he has at least $\frac{1}{2}$ of the pieces of art to choose from. $\frac{66}{2} = 33$ There could be other art pieces priced at $280 above the median so that is why we say "at least".
7	B	To find the most number of eggs laid, add the least number of eggs to the range. The range is $\frac{1}{3}$ of 9 = 3. 9 + 3 = 12 eggs
8	D	The range is the difference between the largest and smallest numbers. To find the range, subtract 48 minutes from 3 hours and 26 minutes. Change 3 hours and 26 minutes to 206 minutes 206 − 48 = 158 minutes Two hours = 120 minutes Subtract 158 − 120 to get 38 minutes remaining That makes 2 hours and 38 minutes
9	C	The range is the difference between the smallest and largest number. The largest number is 174 and the smallest number is 50. Subtract 50 from 174 to find the range. 174 − 50 = 124
10	D	To find the range, subtract the smallest number from the largest number. The largest number is 280. The smallest number is half of 280. $\frac{280}{2} = 140$ 280 − 140 = 140
11		In the first combination, it is given that Lisa has already selected 2 x 5 = 10 paper rolls. Therefore, a = 12 - 10 = 2 In the second combination, it is given that Lisa has already selected (2 x 3) + (1 x 4) = 10 paper rolls. Therefore, b = 12 - 10 = 2 In the third combination, it is given that Lisa has already selected (2 x 2) + (1 x 4) = 8 paper rolls. Therefore, c = 12 - 8 = 4
12		City A = 33.75m City B = 22.50 m City C = 29 m Find the sum of the height of all four buildings and divide by 4 to determine the mean.

Lesson 3: Central Tendency

Question No.	Answer	Detailed Explanations
1	B	To find the average (mean) height of the plants, the heights would first be totaled. Then the total would be divided by 9 (the number of plants in all). 12 + 15 + 11 + 17 + 19 + 21 + 13 + 11 + 16 = 135. 135 divided by 9 equals 15. The average (mean) height is 15 centimeters. To find the median height, the numbers would be arranged in increasing order. The ordered set becomes: {11, 11, 12, 13, 15, 16, 17, 19, 21} The median is the middle value: 15 centimeters.
2	D	To find the average heel height, first figure out how many of each height Stacy has. Create an equation to figure out the number of 4 inch heels that Stacy has. Add 20 + 15 + 20 + x = 60 55 + x = 60 x = 5 Set up an equation to figure out the mean. You need to multiply the number of shoes and the heel height and then add them together and divide by the number of shoes. $$\frac{1(20) + 2(15) + 3(20) + 4(5)}{60} = x$$ x = 2.2 inches
3	D	To find the median, rearrange the numbers in the data set from lowest to highest. {16, −10, 13, −8, −1, 5, 7, 10} −10, −8, −1, 5, 7, 10, 13, 16 Because this set has an even number of terms, add the two middle numbers together and divide by 2. 5 + 7 = 12 12/2 = 6 6 is the median.
4	C	{5, −10, 14, 6, 8, −2, 11, 3, 6} The mode is 6 because it appears most often. To find the median, list the numbers in order from smallest to largest. −10, −2, 3, 5, 6, 6, 8, 11, 14 The median is 6 because it is in the middle.

Question No.	Answer	Detailed Explanations
5	C	0, 2, 10, 14, 15, 19, 25 Median is 14. (A) If 14 is added to the set, median = 14. So, it does not change. (B) If 50 is added to the set, median = $\frac{14+15}{2}$ = 14.5. So, median increases by 0.5. (C) If 5 is added to the set, median = $\frac{10+14}{2}$ = 12. So, median decreases by 2. (D) If 15 is added to the set, median = $\frac{14+15}{2}$ = 14.5. So, median increases by 0.5. Therefore, median would be affected most by adding the score 5 to the set. Option (C) is the correct answer.
6	C	{−7, 18, 29, 4, −3, 11, 22} There is no mode because no numbers repeat To find the median, list the numbers in order from smallest to largest. −7, −3, 4, 11, 18, 22, 29 The median = 11. To find the average, add all of the numbers together and divide by 7. $\frac{-7 + 18 + 29 + 4 + (-3) + 11 + 22}{7} = x$ x = 10.57
7	C	Mean for original set: 1 + 2 + 6 = 9 $\frac{9}{3} = 3$ Median for original set: 2 Mean after adding 3: 1 + 2 + 3 + 6 = 12 $\frac{12}{4} = 3$ Median after adding 3: 2 + 3 = 5 $\frac{5}{2} = 2.5$
8	A	To find the average, add all of the prices together. $0.60 + $0.80 + $0.45 = $1.85 Then, divide by the number of ice cream treats. $\frac{\$1.85}{3} = \$0.616$, or $0.62

Question No.	Answer	Detailed Explanations
9	B	For Marcel to have an average of 35 stamps, he will need to have a total of 175 stamps (5*35). Set up an equation to determine how many stamps the 5th collection should have. Let x represent the 5th collection. $28 + 62 + 12 + 44 + x = 175$ $146 + x = 175$ Subtract 146 from both sides. $146 + x - 146 = 175 - 146$ $x = 29$ stamps
10	B	{67, 29, 40, −12, 88, −7, 11} the mean is 30.857 {67, 29, 40, −12, 88, −7, 11, 53} the mean is 33.625 $33.625 - 30.857 = 2.768$
11	C	First, add the numbers in the set. $25.1 + 19.6 + 88.5 + 0 + -3 + 19.8 = 150$ Then divide the total by the number of terms in the set. $\dfrac{150}{6} = 25$

12

	3	4	8
4, 2, 3, 6, 4, 9, 7		◯	
6, 3, 2, 8, 1, 3, 6	◯		
8, 9, 4, 8, 1, 10, 3			◯

To determine the median number, first, order the set of numbers from least to greatest. The median number will be the number in the center.

13

	11	15	18
12, 18, 11, 20, 20			◯
14, 20, 18, 11, 15		◯	
8, 6, 15, 11, 18	●		

To determine the median number, first, order the set of numbers from least to greatest. The median number will be the number in the center.

Lesson 4: Graphs & Charts

Question No.	Answer	Detailed Explanations
1	A	Using the data to create ordered pairs (x, y), the first choice is the only graph that accurately represents the ordered pairs.
2	C	To find the number of students who passed, add the number of students who scored in the ranges of 61 − 70 (3), 71 − 80 (6), 81−90 (5), and 91 −100 (9). 3 + 6 + 5 + 9 = 23 students
3	D	To find the number of students who scored a 90 or below, add the number of students who scored in the ranges of 51 − 60 (2), 61 − 70 (3), 71 − 80 (6), 81−90 (5). 2 + 3 + 6 + 5 = 16 students
4	D	On Saturday, the greatest range was seen, with a high temperature of 88 and a low temperature of 62, making the range 26 degrees.
5	B	Add together all of the students and then subtract that number from the total number. Jazz: 30 students Concert Band: 42 students Orchestra: 32 students Chorus: 30 students 30 + 42 + 32 + 30 = 134 180 − 134 = 46 students do not participate in any kind of music activity
6	D	Pop and R & B together would make up more than half of the pie chart, or above 50%.
7	B	5 students scored an 81-90 on the test out of 25 students. Convert this fraction into percentage. $\frac{5}{25} \times 100 = 20\%$
8	C	Socks and undergarments together appear to take up more than a tenth, but less than a quarter, of the pie chart. The percentage would be between 10% and 25%.

Question No.	Answer	Detailed Explanations
9	C	Friday and Saturday both had temperatures of 80 degrees or higher. That means that $\frac{2}{7}$ days were 80 degrees or more. To find the percentage, divide 2 by 7. Convert $\frac{2}{7}$ into+ percentage by multiplying $\frac{2}{7}$ with 100. $\frac{2}{7} \times 100 = 28.57\%$ Round to the nearest tenth making the percentage 28.6%
10	A	Jazz Band: 19 + 11 = 30 students Chorus: 12 + 18 = 30 students 60 students participate in either jazz band or chorus. To find the percentage, divide 60 by 132. $60 \div 132 \approx 0.45$ To change the decimal to a percent, move the decimal point to the right two places. 0.45 = 45%
11	A, E, & F	A. 90 children participated in volleyball and 90 children participated in walking. E. 90 children played volleyball and only 80 parents played volleyball. F. 80 parents played volleyball and 80 parents ate lunch.
12	280	Parents volleyball: 80 Children volleyball: 90 Parents lunch: 80 Children lunch: 30 Total 280

Lesson 5: Data Interpretation

Question No.	Answer	Detailed Explanations
1	D	To determine the number of goals he would need, Luis will need a total of 100 (20 * 5) goals. Set up an equation. Let x represent the goals needed for the 5th season: 14 +15 + 18 + 21 + x = 100 72 + x = 100 Subtract 72 from both sides. 72 + x - 72 = 100 x = 28
2	B	Stacy would figure out the mode. The mode will tell her which heel height occurs most often when she lists them out.
3	A	To figure out the percentage, add together all of the ice cream cones sold by stand B. 24 + 8 + 14 = 46 To find the percentage, divide 24 by 46 24/46 = .522 (when rounded to the nearest thousandth) To make the decimal into a percentage, move the decimal point to the right two places and round down. 52%
4	C	To figure out the percentage, add together all of the ice cream cones sold by stand C. 20 + 16 + 13 = 49 To find the percentage, divide 16 by 49 16/49 = .327 (when rounded to the nearest thousandth) To make the decimal into a percentage, move the decimal point to the right two places and round up. 33%
5	A	The number that shows up most often is the mode. .228 is the mode.
6	C	To find the mean, add all of the numbers together and divide by 9. 5 + -10 +14 + 6 +8 +-2 +11 + 3 + 6 = 41 41 ÷ 9 ≈ 4.6 (when rounded to the nearest tenth) 5 + -10 +14 + 6 +8 +-2 +11 + 3 + 6 + 17 = 58 58 ÷ 10 = 5.8 When 17 is added, the mean is 5.8 5.8 - 4.6 = 1.2

Question No.	Answer	Detailed Explanations
7	D	Andrea needs to know the median. To find it, list all of the numbers in order from smallest to largest. 3, 4, 8, 9, 11, 13, 16
8	A	To figure out the percentage, add together all of the ice cream cones sold. Stand C: 20 + 16 + 13 = 49 Stand B: 24 + 8 + 14 = 46 Stand A: 15 + 22 + 10 = 47 49 + 47 + 46 = 142 Twist cones: 10 + 14 + 13 = 37 To find the percentage, divide 37 by 142 37/142 = .261 (when rounded to the nearest thousandth) To make the decimal into a percentage, move the decimal point to the right two places and round down. 26%
9	B	In order to get a representative sample, the person who is conducting the survey needs to get a good sample of what they are surveying. Amy looked at 20 magazines, which is more than half (2/3 to be exact)
10	C	Travis has 114 possible outfits. To figure that out you multiply the number of shirts by the number of khaki pants. 19 * 6 = 114 To find the outfits with blue shirts, multiply 8 by 6 to get 48 To find the percentage, divide 48 by 114. 48/114 = 0.421 (when rounded to the nearest thousandth) To make the decimal into a percentage, move the decimal point to the right two places and round down. 42% Alternate Method : Shirts can be chosen in (4 + 8 + 7) = 19 ways. Blue shirt can be chosen in 8 ways. To find the percentage of outfits with blue shirts, divide 8 by 19 8/19 = 0.421 (when rounded to the nearest thousandth) or 42.1% or 42% (when rounded to the nearest whole number)

11

Travis has 114 possible outfits. To figure that out you multiply the number of shirts by the number of khaki pants. 19 * 6 = 114

To find the outfits without red shirt, add together the blue and white shirts (8 + 7), multiply 15 by 6 to get 90

To find the percentage, divide 90 by 114.

90/114 = .789 (when rounded to the nearest thousandth)

To make the decimal into a percentage, move the decimal point to the right two places and round up.

79%

Alternate Method : Shirts can be chosen in (4 + 8 + 7) = 19 ways. Blue shirt or white shirt (i.e. outfits without red shirt) can be chosen in (8 + 7) = 15 ways.

To find the percentage of outfits without red shirts, divide 15 by 19

15/19 = 0.789 (when rounded to the nearest thousandth) or 78.9% or 79% (when rounded to the nearest whole number)

12

	5	10	20	25	30
0 – <5 cm		⭕			
5 – <10 cm	⭕				
10 – <15 cm		⭕			
15 – <20 cm			⭕		
20 – <25 cm					⭕

13   B, C, & D   B, C, and D would be non-biased.

A would not be a good representation because participants were voluntary, which means those who chose to participate would most likely have similar shopping habits.

E would be biased because it targets a specific age group.

Lesson 6: Describing the Nature

Question No.	Answer	Detailed Explanations
1	D	Use the counting principle to determine the number of combinations if there are 3 types of sandwiches and 4 types of sides for lunch by: $3 * 4 = 12$ There are 12 options.
2	A	{6, 14, 28, 44, 2, −6} the mean is 14.6 {6, 14, 28, 44, 2, −6, −8} the mean is 11.4 The mean will decrease with the addition of the number −8.
3	B	To find the average, add all of the prices together and divide by the number of ingredients, which is 6. $$\frac{\$4.89 + \$2.13 + \$1.10 + \$3.75 + \$0.98 + \$2.46}{6} = \$2.55$$
4	C	To receive an average of 87.0 on the 9 tests, Robert accumulated a total of 783 points. He scored a total of 687 on the first 8 tests. That means his score on the last test was $783 - 687 = 96$.
5	D	Just knowing the median and the range is not enough information to figure out the lowest and highest scores.
6	B	In order to get a representative sample, the person who is conducting the survey needs to get a good sample of what they are surveying. Marcus would need to spend more than 2 hours at the zoo to get a good sample. He should also go on more than one Saturday.
7	B	To figure out how many classmates Carl surveyed, add together all of the numbers in the "Number of Classmates" column. $1 + 4 + 7 + 14 + 2 = 28$
8	C	To figure out the percentage, figure out the total number of quilt squares. $15 + 22 + 8 + 12 + 13 + 10 + 6 = 86$ Figure out how many squares are not flowers or butterflies. $15 + 22 + 12 + 13 + 10 = 72$ To find the percentage, divide the number of squares that are not flowers or butterflies (72) by the total number of squares (86). $$\frac{72}{86} = .837$$ To make the decimal into a percentage, move the decimal point to the right two places and round up. 84%

Question No.	Answer	Detailed Explanations
9	A	To find out how many classmates are taller than 5 feet, add together all of the classmates that are 5' 1" or taller. 7 + 14 + 2 = 23
10	A	Travis has 114 possible outfits. To figure this out multiply the number of shirts by the number of khaki pants. 19 * 6 = 114 To find the outfits without a red shirt, add together the blue and white shirts (8 + 7), multiply 15 by 6 to get 90 To find the percentage, divide 90 by 114. $$\frac{90}{114} = .789$$ To make the decimal into a percentage, move the decimal point to the right two places and round up. 79%
11	A	The numbers added together equal 36, and 36/6 = 6
12	A	Dana's median score is 88.

Lesson 7: Context of Data Gathered

Question No.	Answer	Detailed Explanations
1	B	Find the total number of newspapers leftover and divide by the number of days. $$\frac{3+6+1+0+2+3+6}{7} = \frac{21}{7} = 3 \text{ newspapers}$$
2	D	Find the mean, median, and mode of the data set and compare. Mean: $\frac{13+18+14+12+15+19+11+15}{8} = \frac{117}{8} = 14.625$ Median: 11, 12, 13, 14, 15, 15, 18, 19 → 14.5 Mode: 15 Mode > Mean > Median
3	C	Add the temperatures and divide by the number of temperatures. $$\frac{67+69+69+70+71+72+73+74+74+75}{10} = \frac{714}{10} = 71.4°F$$
4	B	First find the mean. $$\frac{83+92+76+87+89+96+88+91+79+99}{10} = \frac{880}{10} = 88\%$$ Now find the distance between each number and the average. $\|83-88\| = 5$ $\|92-88\| = 4$ $\|76-88\| = 12$ $\|87-88\| = 1$ $\|89-88\| = 1$ $\|96-88\| = 8$ $\|88-88\| = 0$ $\|91-88\| = 3$ $\|79-88\| = 9$ $\|99-88\| = 11$ Take the average of these differences $$\frac{5+4+12+1+1+8+0+3+9+11}{10} = \frac{54}{10} = 5.4\%$$

Question No.	Answer	Detailed Explanations
5	A	First find the mean.

$$\frac{12.3 + 12.6 + 12.2 + 10.9 + 11.3 + 10.3 + 11.7 + 10.7}{8} = \frac{92}{8} = 11.5$$

Now find the distance between each number and the average.

$|12.3-11.5| = 0.8$
$|12.6-11.5| = 1.1$
$|12.2-11.5| = 0.7$
$|10.9-11.5| = 0.6$
$|11.3-11.5| = 0.2$
$|10.3-11.5| = 1.2$
$|11.7-11.5| = 0.2$
$|10.7-11.5| = 0.8$

Take the average of these differences

$$\frac{0.8 + 1.1 + 0.7 + 0.6 + 0.2 + 1.2 + 0.2 + 0.8}{8} = \frac{5.6}{8} = 0.7 min$$

| 6 | A | Put the numbers in order from least to greatest. |

72 82 86 93 95 112 122 130 145
Median = 95

$$Q1 = \frac{82 + 86}{2} = 84$$

$$Q3 = \frac{122 + 130}{2} = 126$$

IQR = 126 − 84 = 42
M = 95, Q1 = 84, Q3 = 126, IQR = 42

| 7 | D | Put the numbers in order from least to greatest. |

$0.75 $3.75 $4.15 $5.20 $6.50
Median = $4.15

$$Q1 = \frac{0.75 + 3.75}{2} = \$2.25$$

$$Q3 = \frac{5.20 + 6.50}{2} = \$5.85$$

IQR = 5.85 - 2.25 = $3.60
M = $4.15, IQR = $3.60

| 8 | B | Find the average of both and then find the difference. |

Anna: $\frac{2.60 + 0 + 7.00 + 4.40 + 3.7}{5} = \frac{17.75}{5} = \$3.55$

Monica: $\frac{5.20 + 6.50 + 3.75 + 0.75 + 4.1}{5} = \frac{20.35}{5} = \$4.07$

Difference: $4.07 − $3.55 = $0.52

Question No.	Answer	Detailed Explanations
9	C	Zahra: 3 8 10 11 12 14 Median $= \frac{10+11}{2} = \frac{21}{2} = 10.5$ Q1 = 8 Q3 = 12 IQR = 12 − 8 = 4 Tyland: 4 6 7 9 12 18 Median $= \frac{7+9}{2} = \frac{16}{2} = 8$ Q1 = 6 Q3 = 12 IQR = 12 − 6 = 6 Tyland has the greater IQR of 6, which is 2 points higher than Zahra's.
10	A	First find the mean. $\frac{20+42+15+23+53+62+58}{7} = \frac{273}{7} = 39$ Now find the distance between each number and the average. $\|20-39\| = 19$ $\|42-39\| = 3$ $\|15-39\| = 24$ $\|23-39\| = 16$ $\|53-39\| = 14$ $\|62-39\| = 23$ $\|58-39\| = 19$ Take the average of these differences $\frac{19+3+24+16+14+23+19}{7} = \frac{118}{7} = 16.9 \; customers$
11	20	To find the median score, order the numbers from least to greatest. 10, 14, 15, 20, 20, 23, 24, 35, 36 Then identify the number that is in the middle. In this list, 20 is in the middle of the list so the median is 20.
12	B	(13 + 19 + 16 + z + 17) / 5 = 16 (65 + z) / 5 = 16 Multiply both sides by 5, 65 + z = 16 x 5 = 80 Subtract 65 from both sides to isolate z z = 80 - 65 = 15

Lesson 8: Relating Data Distributions

Question No.	Answer	Detailed Explanations
1	D	False. There is not enough information to determine the mode score. False. There are 14 out of 36 students who received an 81% or higher. This is less than a half of the students. False. Twelve students received a 70% or lower. True. We know that the median score (the average of the 18th and 19th number) falls between 71% and 80% but we do not know the specific value.
2	A	A) True. The store carries 13 basketball and 5 climbing shoe types for a total of 18 types. This is greater than 15 types, the number of running shoe types. B) False. The graph does not give any information about how much is sold. C) False. The store has 9 types of walking shoes and 10 types of cross trainers and thus has more types of cross trainers than walking shoes. D) False. The graph does not give any information about how much is sold.
3	C	A) False. The weights of Milo's fish have less variability as shown by the closeness of his data points. B) False. The average weight of Milo's fish is $\frac{1.2 + 1.2 + 1.5 + 1.6 + 1.6 + 1.7 + 1.8 + 1.8}{8} = \frac{12.4}{8} = 1.55$ *pounds*. The average weight of Jacque's fish is $\frac{1+1.2+1.2+1.5+1.7+1.8+2+2}{8} = \frac{12.4}{8} = 1.55$ *pounds*. Hence the average weights are the same. C) True. Without any calculations, you can see that Jacque's fish have weights that are farther away from the mean of 1.55 lbs than Milo's fish weight data. D) False. Milo and Jacque caught the same weight in fish. See (B) above.

Question No.	Answer	Detailed Explanations
4	B	A) False. $17 + 20 + 9 + 8 + 12 + 4 = 70$ students

4 — B

A) False. $17 + 20 + 9 + 8 + 12 + 4 = 70$ students

B) True. $\dfrac{[17(11) + 20(12) + 9(13) + 8(14) + 12(15) + 4(16)]}{70} = \dfrac{900}{70} = 12.86$

C) False. The current median age is the average of the 35th and 36th numbers which is $\dfrac{12 + 12}{2} = 12$. If three sixteen year olds arrived the median number would be the 37th number which is 12. Therefore the median age would not change.

D) False. There are 33 students older than the mean of 12.86. Half of 70 would be 35 and 33 is less than 35. Therefore less than half of the students are older than the median.

5 — B

A) False. The best fuel mileage recorded was 38 mpg.

B) True. The Interquartile range is from $27 - 34$ mpg which contains 50% of the data.

C) False. The median fuel mileage was 31 mpg.

D) False. There are twelve data points in all which means that each quartile contains four points and the IQR contains two quartiles or eight points.

6 — D

Note: Each tick mark is 2 inches.

A) False. About 25% of the trees were 7 feet 6 inches or taller.

B) False. Each quartile has the same number of data points.

C) False. The median height is 6 feet 4 inches.

D) True. If Q3 has 15 data points then so do Q1, Q2 and Q4. This makes a total of $15 * 4 = 60$ trees. We say "about" because the median may not be in a quartile.

7 — C

A) False. There are 27 leaves which represent data for 27 students.
B) False. The median number is the 14th number. This is 54.
C) True. The greatest number of jumping jacks performed was 72 and the least was 30. $72 - 30 = 42$.
D) False. There are six leaves in the 3 (30) stem which represents 6 students who performed less than 40 jumping jacks.

Question No.	Answer	Detailed Explanations

8   **D**

A) False The current mean of Tamryn's data is $\dfrac{[2(4) + 4(6) + 2(10) + 2(12)]}{10}$ $= \dfrac{76}{10} = 7.6$. If a 10 were added the new mean would be $\dfrac{[2(4) + 4(6) + 2(10) + 2(12) + 10]}{11} = \dfrac{86}{11} = 7.6$.

B) False. Tamryn: $\dfrac{[2(4) + 4(6) + 2(10) + 2(12)]}{10} = \dfrac{76}{10} = 7.8$. Clio: $\dfrac{[2(6) + 2(7) + 3(8) + 9 + 2(10)]}{10} = \dfrac{79}{10} = 7.9$. Clio used more beads than Tamryn on the average.

C) False. As shown by the spread of the data, Tamryn's data has more variability than Clio's.

D) True. The median of Clio's data is 8. If Clio added another 8 the median would remain the same.

9   **A**

A) True. There are a total of $2 + 5 + 18 + 4 + 1 = 30$ days. 18 of these days had a temperature of $70 - 74°F$. This is more than half of the days.

B) False. $4 + 1 = 5$ days had temperatures above $74°F$. This is $\dfrac{5}{30}$ $= 16.6\%$ which is less than 20%.

C) False. There is no way to determine the average temperature without specific data points.

D) False. There were $5 + 2 = 7$ days below $74°F$ and $4 + 1 = 5$ days above $74°F$. Therefore there were more days below $74°F$ than days above $74°F$.

10   **B**

A) False. The mean is

$$\dfrac{4 + 6 + 9 + 10 + 13 + 14 + 17 + 17 + 17 + 18 + 20 + 20 + 22 + 22 + 25 + 31 + 34}{17} = \dfrac{299}{17}$$

$= 17.59$. The median is 17. Therefore the mean is greater than the median.

B) True. The middle number is 17.
C) False. The range of fireflies is $34 - 4 = 30$.
D) False. Marcel caught fireflies on 17 nights.

Question No.	Answer		Detailed Explanations

11   Box 3

-
-
-
-
-
-

3

Because box 3 shows that 6 students in this class have 3 siblings. A smaller number of students have 0, 1, 2, 4, 5, or 6 siblings in this class.

12

	10	14	17
Team 1: 15, 16, t, 17, 12, Mode = 17, What is t?			O
Team 2: 7, 14, r, 16, 13 Mean = 12, What is r?	O		
Team 3: 11, 7, 19, 14, z, Median = 14, What is z?		O	

Team 1: t =17 would be the mode because it would be listed two times.
Team 2: (7 + 14 + r + 16 + 13) / 5 = 12
           (50 + r) = 12 x 5 = 60
           r = 60 - 50 = 10
Team 3: Arrange the known numbers in ascending order : 7, 11, 14, 19. It is given median is 14. Therefore, the unknown number z must be more than or equal to 14.
z = 14

Additional Information

Test Taking Tips

1) **The day before the test,** make sure you get a good night's sleep.

2) **On the day of the test,** be sure to eat a good hearty breakfast! Also, be sure to arrive at school on time.

3) **During the test:**

- **Read every question carefully.**

    - Do not spend too much time on any one question. Work steadily through all questions in the section.
    - Attempt all of the questions even if you are not sure of some answers.
    - If you run into a difficult question, eliminate as many choices as you can and then pick the best one from the remaining choices. Intelligent guessing will help you increase your score.
    - Also, mark the question so that if you have extra time, you can return to it after you reach the end of the section.
    - Some questions may refer to a graph, chart, or other kind of picture. Carefully review the graphic before answering the question.
    - Be sure to include explanations for your written responses and show all work.

- **While Answering Multiple-Choice (EBSR) questions.**

    - Select the bubble corresponding to your answer choice.
    - Read **all** of the answer choices, even if think you have found the correct answer.

- **While Answering TECR questions.**

    - Read the directions of each question. Some might ask you to drag something, others to select, and still others to highlight. Follow all instructions of the question (or questions if it is in multiple parts)

Frequently Asked Questions(FAQs)

For more information on the assessment, visit
www.lumoslearning.com/a/ksa-faqs
OR Scan the **QR Code**

What if I buy more than one Lumos tedBook?

Step 1 → **Visit the link given below and login to your parent/teacher account**
www.lumoslearning.com

Step 2 → **For Parent**
Click on the horizontal lines (☰) in the top right-hand corner and select **"My tedBooks"**. Place the Book Access Code and submit.

For Teacher
Click on "My Subscription" under the "My Account" menu in the left-hand side and select **"My tedBooks"**. Place the Book Access Code and submit.

Note: See the first page for access code.

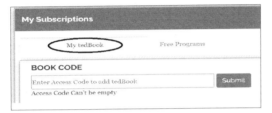

Step 3 → **Add the new book**
To add the new book for a registered student, choose the **'Student'** button and click on submit.

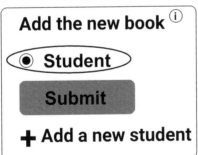

To add the new book for a new student, choose the **'Add New Student'** button and complete the student registration.

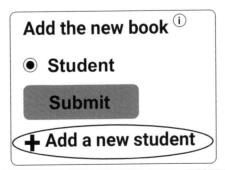

Progress Chart

Standard	Lesson	Score	Date of Completion
KAS			
KY.6.RP.1	Expressing Ratios		
KY.6.RP.2	Unit Rates		
KY.6.RP.3	Solving Real World Ratio Problems		
KY.6.RP.3.B	Solving Unit Rate Problems		
KY.6.RP.3.C	Finding Percent		
KY.6.RP.3.D	Measurement Conversion		
KY.6.NS.1	Division of Fractions		
KY.6.NS.2	Division of Whole Numbers		
KY.6.NS.3	Operations with Decimals		
KY.6.NS.4	Using Common Factors		
KY.6.NS.5	Positive and Negative Numbers		
KY.6.NS.6.A	Representing Negative Numbers		
KY.6.NS.6.B	Ordered Pairs		
KY.6.NS.6.C	Number Line & Coordinate Plane		
KY.6.NS.7	Absolute Value		
KY.6.NS.7.B	Rational Numbers in Context		
KY.6.NS.7.C	Interpreting Absolute Value		
KY.6.NS.7.D	Comparisons of Absolute Value		
KY.6.NS.8	Coordinate Plane		
KY.6.EE.1	Whole Number Exponents		
KY.6.EE.2.A	Expressions Involving Variables		
KY.6.EE.2.B	Identifying Expression Parts		
KY.6.EE.2.C	Evaluating Expressions		
KY.6.EE.3	Writing Equivalent Expressions		
KY.6.EE.4	Identifying Equivalent Expressions		
KY.6.EE.5	Equations and Inequalities		
KY.6.EE.6	Modeling with Expressions		
KY.6.EE.7	Solving One-Step Equations		
KY.6.EE.8	Representing Inequalities		
KY.6.EE.9	Quantitative Relationships		

Standard	Lesson	Score	Date of Completion
KAS			
KY.6.G.1	Area		
KY.6.G.2	Surface Area and Volume		
KY.6.G.3	Coordinate Geometry		
KY.6.G.4	Nets		
KY.6.SP.1	Statistical Questions		
KY.6.SP.2	Distribution		
KY.6.SP.3	Central Tendency		
KY.6.SP.4	Graphs & Charts		
KY.6.SP.5	Data Interpretation		
KY.6.SP.5.B	Describing the Nature		
KY.6.SP.5.C	Context of Data Gathered		
KY.6.SP.5.D	Relating Data Distributions		

Grade **6**

Lumos Learning
Step Up Your Skills

KENTUCKY
ENGLISH
LANGUAGE ARTS LITERACY
KSA Practice

(((tedBook)))
ONLINE

2 Practice Tests

Personalized Study Plan

ELA Strands   Literature • Informational Text • Language

Available

• At Leading book stores

• Online www.LumosLearning.com

Made in the USA
Columbia, SC
24 March 2025